Lecture Notes in Physics

Founding Editors

Wolf Beiglböck

Jürgen Ehlers

Klaus Hepp

Hans-Arwed Weidenmüller

Volume 1044

Series Editors

Roberta Citro, Salerno, Italy

Peter Hänggi, Augsburg, Germany

Betti Hartmann , London, UK

Morten Hjorth-Jensen, Oslo, Norway

Maciej Lewenstein, Barcelona, Spain

Satya N. Majumdar, Orsay, France

Luciano Rezzolla, Frankfurt am Main, Germany

Angel Rubio, Hamburg, Germany

Wolfgang Schleich, Ulm, Germany

Stefan Theisen, Potsdam, Germany

James D. Wells, Ann Arbor, MI, USA

Gary P. Zank, Huntsville, AL, USA

The series Lecture Notes in Physics (LNP), founded in 1969, reports new developments in physics research and teaching - quickly and informally, but with a high quality and the explicit aim to summarize and communicate current knowledge in an accessible way. Books published in this series are conceived as bridging material between advanced graduate textbooks and the forefront of research and to serve three purposes:

- to be a compact and modern up-to-date source of reference on a well-defined topic;
- to serve as an accessible introduction to the field to postgraduate students and non-specialist researchers from related areas;
- to be a source of advanced teaching material for specialized seminars, courses and schools.

Both monographs and multi-author volumes will be considered for publication. Edited volumes should however consist of a very limited number of contributions only. Proceedings will not be considered for LNP.

Volumes published in LNP are disseminated both in print and in electronic formats, the electronic archive being available at springerlink.com. The series content is indexed, abstracted and referenced by many abstracting and information services, bibliographic networks, subscription agencies, library networks, and consortia.

Proposals should be sent to a member of the Editorial Board, or directly to the responsible editor at Springer:

Dr Lisa Scalone
lisa.scalone@springernature.com

Philip K. Schwartz

Newton–Cartan Gravity

A Modern Introduction to Geometrised Newtonian Gravity

Springer

Philip K. Schwartz ⓘ
Institute of Theoretical Physics
Leibniz University Hannover
Hannover, Germany

ISSN 0075-8450 ISSN 1616-6361 (electronic)
Lecture Notes in Physics
ISBN 978-3-032-03966-8 ISBN 978-3-032-03967-5 (eBook)
https://doi.org/10.1007/978-3-032-03967-5

This Springer imprint is published by the registered company Springer Nature Switzerland AG
The registered company address is: Gewerbestrasse 11, 6330 Cham, Switzerland

Preface

Newton–Cartan gravity is a reformulation of Newtonian gravity in differential-geometric language, bringing it closer to general relativity (GR) than the standard formulation. This allows for a coordinate-free understanding of Newtonian physics from a spacetime point of view, and of how GR reduces to Newtonian gravity in the Newtonian limit[1].

This book grew out of a lecture course on Newton–Cartan gravity that I taught several times at Leibniz University Hannover. It develops Newton–Cartan gravity from the ground up, taking a modern point of view. Its aim is to provide a solid introduction into the subject for both researchers and graduate students in gravitational physics and nearby subjects, offering a route towards the current research literature.

The book begins with a brief historical introduction in Chap. 1. Chapter 2 introduces and develops Galilei geometry, which underlies Newton–Cartan gravity. In Chap. 3, this foundation is used to discuss 'classical' Newton–Cartan gravity: first, the theory is axiomatised as a spacetime theory similar to GR, and some basic kinematical notions are analysed in its context. Next, the recovery of standard Newtonian gravity from Newton–Cartan gravity is discussed in detail. Finally, the Newtonian limit of Lorentzian geometry and GR to Galilei geometry and Newton–Cartan gravity is developed.

In the discussion of Galilei manifolds in Chap. 2, I treat the general case of Galilei connections *with possibly non-vanishing torsion* whenever possible and sensible. For most of the classical discussion in Chap. 3, only the case of torsion-free Galilei connections is necessary. However, since torsion is often allowed for or even necessary in modern applications of Newton–Cartan gravity, I decided to cover the general case.

---

[1] As a matter of principle, I am going to avoid the term 'non-relativistic' when speaking about Newtonian limits whenever possible. The reason for that is that Newtonian physics is relativistic as well, in the sense of satisfying the principle of relativity. The principle is just implemented differently than in special or general relativity (or their modifications), namely by the Galilei instead of the Poincaré (inhomogeneous Lorentz) group.

The second part of the book (Chaps. 4 and 5) deals with a somewhat gauge-theoretic perspective on Newton–Cartan theory, which is an important ingredient in many of its modern applications. This perspective is based on the realisation that Galilei geometry is intimately related to the Bargmann algebra, the one-dimensional central extension of the inhomogeneous Galilei algebra. In Chaps. 4 and 5, this formalism is presented from a global, principal-bundle point of view, and its applications are discussed.

Recently (starting in the early 2010s), generalisations of Newton–Cartan gravity have found novel applications in condensed matter physics, in research on aspects of quantum gravity, such as holography and 'non-relativistic' limits of string theory, and in the coordinate-free description of the post-Newtonian expansion of GR. Chapter 6 provides an outlook on torsional Newton–Cartan gravity, with a particular focus on aspects of the post-Newtonian expansion, as well as on basics of string Newton–Cartan geometry.

Apart from the main text, the book contains three appendices: Appendix A briefly discusses semidirect products of Lie groups and semidirect sums of Lie algebras; and Appendices B and C contain details of mathematical constructions used in Chaps. 5 and 6.

The book is written in a fully rigorous mathematical style. Where appropriate, I make full use of differential-geometric concepts and techniques, in order to guide the reader towards a truly geometric understanding of the theory. Therefore, a good understanding of basic differential geometry is indispensable to follow the presentation. For Chaps. 4 and 5, knowledge of principal bundles and associated vector bundles is also necessary.

The philosopher of physics David Malament has written a textbook on foundational issues of GR, *Topics in the Foundations of General Relativity and Newtonian Gravitation Theory* (Chicago Lectures in Physics, The University of Chicago Press, 2012), with an extensive chapter on Newton–Cartan gravity. This is, in my opinion, an excellent presentation of the subject's classical aspects, as discussed in the present book in Chaps. 2 and 3 (except for the Newtonian limit of GR, which is unfortunately not covered in Malament's book). For readers who want a different perspective on topics treated in these chapters, Malament's book is the go-to resource. Note, however, that it is aimed at students with no prior exposure to GR or differential geometry, and therefore often has to proceed in a rather elementary way, thereby somewhat precluding appreciation of some geometric aspects.

I am indebted to several people who were in one way or another involved in the writing of this book. First I wish to thank Domenico Giulini for supporting my idea of teaching a course on Newton–Cartan gravity. Further, I thank the students taking the various incarnations of this course for their interest. Without these lectures, the book would never have existed. I also thank Nico as well as Arian von Blanckenburg for a lot of valuable discussions on the various topics covered in the book. Finally, Arian deserves special thanks for proofreading the entire manuscript. All remaining errors are, of course, my own fault.

Hannover, Germany   Philip K. Schwartz
June 2025

Notation and Conventions

Here, we list our basic notation and conventions.

Manifolds

All differentiable manifolds are assumed smooth, i.e. C^∞. For the sake of simplicity, all objects on manifolds are also assumed smooth, unless otherwise stated. However, in almost all cases, a finite degree of differentiability is sufficient.

Regarding topological assumptions, we use the standard conventions of modern differential topology, i.e. manifolds are always Hausdorff and second countable/paracompact.

We denote the space of (smooth) sections of a fibre bundle $E \to M$ by $\Gamma(E)$; the space of local sections on an open subset $U \subset M$ is $\Gamma(U, E) := \Gamma(E|_U)$. For the space of differential k-forms on some manifold M, we use the standard notation $\Omega^k(M) = \Gamma(\bigwedge^k T^*M)$. The k-fold symmetric tensor power of a vector space (or bundle) V we denote by $\bigvee^k V$; so e.g. $\Gamma(\bigvee^2 T^*M)$ is the space of symmetric covariant two-tensor fields on M.

Wedge Products

Let V be a vector space. When understanding elements of the exterior power $\bigwedge^k V^*$ as alternating k-linear forms on V, we use the convention that the application of a wedge product of one-forms $\alpha^1, \ldots, \alpha^k \in V^*$ to vectors $v_1, \ldots, v_k \in V$ is given by[2]

$$(\alpha^1 \wedge \cdots \wedge \alpha^k)(v_1, \ldots, v_k) = \det(\alpha^i(v_j))_{i,j=1}^k. \tag{1}$$

---

[2] This is the majority convention in modern differential geometry. However, it *differs* from the convention used in the influential foundational monograph *Foundations of Differential Geometry* by Kobayashi and Nomizu: they have an additional prefactor of $\frac{1}{k!}$ on the right-hand side of (1), and hence *no* prefactor on the right-hand side of (2).

This implies that the wedge product of $\alpha \in \bigwedge^k V^*$, $\beta \in \bigwedge^l V^*$ is given by

$$\alpha \wedge \beta = \frac{(k+l)!}{k!\,l!} \mathrm{Alt}(\alpha \otimes \beta), \tag{2}$$

where Alt denotes total antisymmetrisation of multilinear forms.

Lie Groups and Algebras

For a Lie group denoted by a Latin capital letter (or by several letters), the corresponding Lie algebra is denoted by the corresponding lowercase Fraktur letter, e.g. $\mathfrak{g} = \mathrm{Lie}(G)$.

The semidirect product of Lie groups H and N with respect to a homomorphism $\rho : H \to \mathrm{Aut}(N)$ is denoted by $H \ltimes_\rho N$; the semidirect sum of Lie algebras $\mathfrak{h}$ and $\mathfrak{n}$ with respect to a homomorphism $\tilde{\rho} : \mathfrak{h} \to \mathrm{Der}(\mathfrak{n})$ is denoted by $\mathfrak{h} \oplus_{\tilde{\rho}} \mathfrak{n}$. In the case of both semidirect products and sums, if the homomorphism is clear from context we omit it from the notation. A discussion of the basics of semidirect products of Lie groups and semidirect sums of Lie algebras may be found in Appendix A.

Forms on Principal Bundles

Forms living on the total space of a principal bundle are denoted by boldface letters. Their local representatives on the base manifold, i.e. pullbacks along local sections of the bundle, are denoted by the corresponding non-boldface letters, with the local section being understood from context. For example, if $\boldsymbol{\omega} \in \Omega^1(P, \mathfrak{g})$ is a connection form on a principal G-bundle $P \xrightarrow{\pi} M$, then for a local section $\sigma \in \Gamma(U, P)$ on an open set $U \subset M$ the corresponding local connection form is $\omega = \sigma^* \boldsymbol{\omega} \in \Omega^1(U, \mathfrak{g})$.

Given a representation $\rho : G \to \mathrm{GL}(V)$, the space of ρ-tensorial k-forms on the total space P of a principal G-bundle is denoted by $\Omega^k_\rho(P, V)$ or, if the representation is clear from context, by $\Omega^k_G(P, V)$.

Index Notation

In index notation, the Einstein summation convention is used throughout. As generic coordinate indices, we use lowercase Greek letters from the middle of the alphabet, unless otherwise stated. When a local coordinate chart (x^μ) on a manifold M is chosen, we denote the induced coordinate vector fields by

$$\partial_\mu := \frac{\partial}{\partial x^\mu}. \tag{3}$$

For example, the coordinate component decomposition of a vector field reads $X = X^\mu \partial_\mu$.

For total symmetrisation and antisymmetrisation of tensors, we use the common notational conventions of enclosing the respective indices in round or square brackets, respectively; e.g.

$$X_{(\mu\nu)} = \frac{1}{2}(X_{\mu\nu} + X_{\nu\mu}), \tag{4a}$$

$$X^{[\mu}{}_\rho{}^{\nu]} = \frac{1}{2}(X^\mu{}_\rho{}^\nu - X^\nu{}_\rho{}^\mu). \tag{4b}$$

Enclosed indices which are *not* to be (anti-)symmetrised are surrounded by straight lines, such that for example in the expression $X^{(\mu\nu}h^{|\rho|\sigma)}$, only μ, ν, σ are symmetrised.

For example, given differential forms $\alpha \in \Omega^k(M)$, $\beta \in \Omega^l(M)$, the coordinate component expression for their wedge product is

$$(\alpha \wedge \beta)_{\mu_1 \ldots \mu_{k+l}} = \frac{(k+l)!}{k!\,l!} \alpha_{[\mu_1 \ldots \mu_k} \beta_{\mu_{k+1} \ldots \mu_{k+l}]}, \tag{5}$$

and that for the exterior derivative of α is

$$(\mathrm{d}\alpha)_{\mu_1 \ldots \mu_{k+1}} = (k+1)\partial_{[\mu_1}\alpha_{\mu_2 \ldots \mu_{k+1}]}. \tag{6}$$

Linear Connections

By a *linear connection* on a manifold M we mean a covariant derivative operator ∇ on the tangent bundle TM, which we extend to tensor bundles in the natural way, i.e. via demanding a Leibniz rule. For example, for vector fields $X, Y \in \Gamma(TM)$ and a one-form $\alpha \in \Omega^1(M) = \Gamma(T^*M)$, we have

$$(\nabla_X \alpha)(Y) = X(\alpha(Y)) - \alpha(\nabla_X Y). \tag{7}$$

The torsion of the connection is denoted by T and the curvature tensor by R, i.e.

$$T(X, Y) := \nabla_X Y - \nabla_Y X - [X, Y], \tag{8a}$$

$$R(X, Y)Z := \nabla_X \nabla_Y Z - \nabla_Y \nabla_X Z - \nabla_{[X,Y]}Z \tag{8b}$$

for vector fields $X, Y, Z \in \Gamma(TM)$. In components, we use the usual index conventions for torsion and curvature, namely

$$T(X, Y) = T^\rho{}_{\mu\nu} X^\mu Y^\nu \partial_\rho, \tag{9a}$$

$$R(X, Y)Z = R^\mu{}_{\nu\rho\sigma} X^\rho Y^\sigma Z^\nu \partial_\mu. \tag{9b}$$

Connection coefficients in local coordinates are defined in the usual way, i.e. by expanding the covariant derivatives of coordinate vector fields according to

$$\nabla_{\partial_\mu} \partial_\nu = \Gamma^\rho{}_{\mu\nu} \partial_\rho. \tag{10}$$

Note that the *first* lower index is the 'form index'/differentiation index, such that the coordinate component form of the torsion is

$$T^\rho{}_{\mu\nu} = 2\Gamma^\rho{}_{[\mu\nu]}. \tag{11}$$

Lie Derivatives

The Lie derivative of a tensor field $A \in \Gamma(\bigotimes^s TM \otimes \bigotimes^t T^*M)$ with respect to a vector field X is denoted by $\mathcal{L}_X A$.

We make free use of the fact that in coordinates, the Lie derivative satisfies

$$(\mathcal{L}_{\partial_\rho} A)^{\mu_1 \dots \mu_s}_{\nu_1 \dots \nu_t} = \partial_\rho A^{\mu_1 \dots \mu_s}_{\nu_1 \dots \nu_t}, \tag{12}$$

i.e. that the components of a Lie derivative along a coordinate vector field are given by the partial derivatives of the component functions.

Lorentzian Geometry

Our signature convention for Lorentzian metrics is mostly plus, i.e. $(- + \dots +)$.

Contents

Symbols

Here we list symbols that are recurrently used with specific meanings. Note
that some of them are also used as generic variable names. Note further that
Sects. 5.6 and 6.2 introduce specific notations which are only used in these sections
and therefore not listed here.

Text Organisation Symbols

$\square$ End of proof
$\triangle$ End of 'construction-like' item (construction, example, exercise, interpreta-
tion, remark)
✾   End of 'definition-like' item (axioms, definition, notation)

Indices

$\mu, \nu, \rho, \sigma, \ldots$	Coordinate indices
$A, B, C, \ldots$	Basis/frame indices running through the whole range
$a, b, c, \ldots$	Spatial basis/frame indices
t	Temporal basis/frame index on a Galilei manifold
0	Temporal basis/frame index on a Lorentzian manifold

Other Symbols

a	Bargmann form
a	Local representative of a Bargmann form
$B(M)$	Bargmann extension of the Galilei frame bundle
Barg	Bargmann group
$\mathfrak{barg}$	Bargmann algebra
c	Speed of light
D	Differential of a smooth map

d	Exterior derivative
(E_A)	Lorentzian orthonormal basis/frame
(e_A)	Galilei basis/frame
$F(M)$	Linear frame bundle of the manifold M
f	Mass torsion of a Galilei connection with respect to a Bargmann form
f	Local representative of the mass torsion
G	Gravitational constant
$G(M)$	Galilei frame bundle
g	Lorentzian metric
$g^{(0)}$	Order-c^0 coefficient of the Lorentzian metric in the Newtonian limit
Gal	(Homogeneous) Galilei group
$\mathfrak{gal}$	(Homogeneous) Galilei algebra
$\mathsf{GL}(n)$	General linear group in n dimensions
$\mathfrak{gl}(n)$	Lie algebra of real $n \times n$ matrices
h	Space metric
$h^{\mu\nu}$	Components of the space metric
$^{(n)}h$	Metric induced on spacelike vectors
$\underset{v}{h}$	Covariant space metric with respect to the unit timelike vector field v
$h_{\mu\nu}$	Components of the covariant space metric
k	Boost parameter of a Galilei transformation
$\mathcal{L}$	Lie derivative
(M, τ, h)	Galilei manifold
m	Order-c^{-2} coefficient of the inverse Lorentzian metric in the Newtonian limit
n	Dimension of space
$\mathrm{O}(c^{-k})$	Order-k terms in the c^{-1} expansion describing the Newtonian limit
$\mathrm{O}(n)$	Orthogonal group in n dimensions
P	Spatial projector along a unit timelike vector field
R	Rotation parameter of a Galilei transformation; curvature tensor (abstractly)
$R^{\mu}{}_{\nu\rho\sigma}$	Components of the curvature tensor
$R_{\mu\nu}$	Components of the Ricci tensor
$^{(n)}R$	Curvature tensor of spatial leaves
Ric	Ricci tensor (abstractly)
$^{(n)}\mathrm{Ric}$	Ricci tensor of spatial leaves
T_pM	Tangent space to M at $p \in M$
T	Torsion of a linear connection
t	Absolute time coordinate
v	Unit timelike vector field

$\hat{v}$	Unit timelike vector field induced by c^{-1} expansion of a Lorentzian metric; boost-invariant unit timelike vector field determined by a Bargmann form
w	Milne boost vector field
α	Acceleration of a unit timelike vector field
$\Gamma(E)$	Sections of the fibre bundle E
∇	Galilei connection
$\overset{v}{\nabla}$	Special Galilei connection with respect to the unit timelike vector field v
$\overset{(n)}{\nabla}$	Connection induced on spacelike vectors
$\Gamma^{\rho}_{\mu\nu}$	Coordinate connection coefficients
$\overset{v}{\Gamma}{}^{\rho}_{\mu\nu}$	Connection coefficients of the special Galilei connection with respect to v
γ	Curve (worldline) in a manifold
$\delta^{\mu}_{\nu}, \delta^{A}_{B}, \delta^{ab}, \delta_{ab}$	Kronecker delta
$^{(n)}\delta$	Codifferential of spatial leaves
η_{AB}	Components of the Minkowski metric in Lorentzian coordinates
$\boldsymbol{\theta}$	Tensorial one-form on a reduction of the linear frame bundle corresponding to the canonical solder form
θ	Expansion of a unit timelike vector field; canonical solder form
ϖ	(See at ω)
ρ	Mass density; homomorphism from the definition of the Bargmann group
Σ	Spatial leaf
σ	Shear of a unit timelike vector field; local section of a principal bundle
τ	Clock form
ϕ	Newtonian gravitational potential
$\hat{\phi}$	Boost-invariant potential function determined by a Bargmann form
φ	Diffeomorphism
Ω	Newton–Coriolis form of a Galilei connection with respect to a unit timelike vector field
$\Omega^{k}(M)$	Space of differential k-forms on M
$\Omega^{k}_{\rho}(P, V)$	Space of ρ-tensorial k-forms on P
$\boldsymbol{\omega}$	Connection on a principal bundle
ω	Twist of a unit timelike vector field; local connection form
$\boldsymbol{\varpi}^{a}$	Boost-valued part of the principal connection corresponding to a Galilei connection
ϖ^{a}	Boost-valued part of the local connection formof a Galilei connection
∇	(See at Γ)

$\ltimes$	Semidirect product of (Lie) groups
$\oplus$	Semidirect sum of Lie algebras
$[\cdot \wedge \cdot]$	Commutator–wedge product of Lie algebra-valued differential forms

Introduction

1

Abstract

Here, we give a brief historical overview over the development of Newton–Cartan gravity.

The geometrised formulation of Newtonian gravity that is today known as Newton–Cartan gravity goes back to a long article on differential geometry and general relativity (GR) written in 1923 by Élie Cartan [4–6] (who is, of course, also responsible for the development of many essential notions of modern differential geometry, for example the concept of differential forms). In a rather short part of this article, Cartan showed that, in great parallel to GR, Newtonian gravity can be given a truly geometric, coordinate-free formulation, with the Newtonian gravitational force becoming a fictitious force in the process, on the same footing as 'inertial forces'.

In 1926 (published in 1928), Kurt Friedrichs re-discovered the geometric formulation of Newtonian gravity independently [15]. In this article, Friedrichs also discussed for the first time the limiting process from GR to Newtonian gravity in invariant geometric terms.

Important further contributions to Newton–Cartan gravity were made by several researchers in the 1960s and 1970s (for more details on the history, we refer to the bibliographies of the references cited in the following, particularly the articles by Künzle and the book by Malament). First came Andrzej Trautman, who developed the formalism into a fully-fledged theory in a 1963 article on the subject [25] as well as in a section of a 1965 lecture series on GR [26]. Trautman also realised the importance of imposing an additional geometric condition on the curvature tensor in order to recover Newtonian gravity proper from Newton–Cartan gravity. Dombrowski and Horneffer in 1964 [8] were the first to analyse the geometry underlying the theory from a modern differential-geometric point of view, naming it Galilei geometry. In the 1970s, Hans Peter Künzle gave the first complete presentation of the theory in its modern form [21,22], including a disussion of its geometric foundations in precise terms and the first fully coordinate-free presentation of its relation to GR. In these

P. K. Schwartz, *Newton–Cartan Gravity*, Lecture Notes in Physics 1044,
https://doi.org/10.1007/978-3-032-03967-5_1

articles, Künzle was also the first to discuss Galilei geometry from a principal bundle point of view.

In a 1981 article [13,14], Jürgen Ehlers introduced 'frame theory', a framework comprising both GR and Newton–Cartan gravity, and made precise some of the earlier results of Künzle on the relation of GR and Newtonian gravity in its Newton–Cartan formulation. This 'classical' Newton–Cartan gravity as discussed by Künzle and Ehlers is the subject of David Malament's presentation of the theory in Chap. 4 of his 2012 textbook on foundational aspects of GR [23], and it is our object of discussion in Chap. 3 of the present book.

The relation between Galilei geometry and the Bargmann algebra, the one-dimensional central extension of the inhomogeneous Galilei algebra, was first discovered in 1976 (published in 1978) by Christian Duval and Künzle in the context of a geometric description of matter coupling to Newton–Cartan gravity [11]. Soon after, in 1977, Duval and Künzle showed that this perspective enables a global, principal-bundle description of so-called Newtonian connections (the connections describing inertial motion in Newton–Cartan gravity) [10]. In 1983 (published in 1984), they used this formalism to develop a geometric formulation of the Schrödinger equation coupled to Newton–Cartan gravity [12]. This lead to the discovery in 1984 (published in 1985) by Duval, Burdet, Künzle, and Perrin [9] that Galilei geometry arises from Lorentzian geometry not only by a Newtonian limit, but also by so-called null reduction of higher-dimensional Lorentzian geometry, i.e. quotienting by a lightlike symmetry. In 1994 (published in 1995), null reduction was further developed in a more general context by Julia and Nicolai [20].

Newton–Cartan gravity and certain generalisations thereof became a hot topic of research again in the early 2010s, and have stayed so ever since. On the one hand, this is due to the discovery that they can be employed in the effective description of effects in Galilei-relativistic condensed matter physics. This line of research started with an application of Galilei geometry to the quantum Hall effect by Son in 2013 [24]. Subsequently, this led to a broad interest in the condensed matter community; see, e.g., the articles [17,18] and references therein.

On the other hand, modifications of Newton–Cartan gravity have been employed in research on several aspects of quantum gravity. This started with a revival of the Bargmann algebra approach to Galilei geometry in 2010 (published in 2011) by Andringa, Bergshoeff, Panda, and de Roo [2], a group of high energy physicists with a supergravity and string theory background. In 2012, this lead to the construction of string Newton–Cartan geometry by Andringa, Bergshoeff, Gomis, and de Roo [1], giving rise to a broad field of research on applications of modified Galilei geometry in string theory. Seperate from this development, in 2013 (published in 2014) Christensen, Hartong, Obers, and Rollier discovered so-called torsional Newton–Cartan gravity in the context of holography [7]. This, again, spawned a lot of research, with connections also to applications in condensed matter physics and string theory. All of these recent developments have since become deeply interrelated—see, e.g., the 2015 article [16] by Geracie, Prabhu, and Roberts for one of the first explicit connections between the condensed matter, Bargmann, and torsional Newton–Cartan stories from a somewhat mathematical point of view.

Most recently, torsional Newton–Cartan gravity has found applications in 'strong gravity' generalisations of the geometric coordinate-free description of the Newtonian limit of GR, as discussed by Van den Bleeken in 2017 [3], and in the coordinate-free description of the post-Newtonian expansion of GR, starting with Hansen, Hartong, and Obers in 2018 (published in 2019) [19]. For more references on torsional Newton–Cartan gravity, its application in post-Newtonian expansions, and string Newton–Cartan geometry, we refer to the outlook in Chap. 6.

References

1. Andringa, R., Bergshoeff, E., Gomis, J., de Roo, M.: 'Stringy' Newton–Cartan gravity. Class. Quantum Gravity **29**(23), 235020 (2012). https://doi.org/10.1088/0264-9381/29/23/235020
2. Andringa, R., Bergshoeff, E., Panda, S., de Roo, M.: Newtonian gravity and the Bargmann algebra. Class. Quantum Gravity **28**(10), 105011 (2011). https://doi.org/10.1088/0264-9381/28/10/105011
3. Van den Bleeken, D.: Torsional Newton–Cartan gravity from the large c expansion of general relativity. Class. Quantum Gravity **34**(18), 185004 (2017). https://doi.org/10.1088/1361-6382/aa83d4
4. Cartan, E.: Sur les variétés à connexion affine et la théorie de la relativité généralisée (première partie). Ann. Sci. Éc. Norm. Supér. **40**, 325–412 (1923). https://doi.org/10.24033/asens.751. English translation available as [6]
5. Cartan, E.: Sur les variétés à connexion affine et la théorie de la relativité généralisée (première partie) (Suite). Ann. Sci. Éc. Norm. Supér. **41**, 1–25 (1924). https://doi.org/10.24033/asens.753. English translation available as [6]
6. Cartan, E.: On Manifolds with an Affine Connection and the Theory of General Relativity. In: Monographs and Textbooks in Physical Science. Bibliopolis, Napoli (1986). https://bibliopolis.it/shop/on-manifolds-with-an-affine-connection-and-the-theory-of-general-relativity/. English translation of original article series [4, 5], translated by Magnon, A., Ashtekar, A
7. Christensen, M.H., Hartong, J., Obers, N.A., Rollier, B.: Torsional Newton-Cartan geometry and Lifshitz holography. Phys. Rev. D **89**(6), 061901 (2014). https://doi.org/10.1103/PhysRevD.89.061901
8. Dombrowski, H.D., Horneffer, K.: Die Differentialgeometrie des Galileischen Relativitätsprinzips. Math. Z. **86**(4), 291–311 (1964). https://doi.org/10.1007/BF01110404
9. Duval, C., Burdet, G., Künzle, H.P., Perrin, M.: Bargmann structures and Newton-Cartan theory. Phys. Rev. D **31**(8), 1841–1853 (1985). https://doi.org/10.1103/PhysRevD.31.1841
10. Duval, C., Künzle, H.P.: Sur les connexions newtoniennes et l'extension non triviale du groupe de Galilée. C. R. Acad. Sci. A **285**, 813–816 (1977). https://www.researchgate.net/publication/267065597_Sur_les_connexions_newtoniennes_et_l%27extension_non_triviale_du_groupe_de_Galilee
11. Duval, C., Künzle, H.P.: Dynamics of continua and particles from general covariance of Newtonian gravitation theory. Rep. Math. Phys. **13**(3), 351–368 (1978). https://doi.org/10.1016/0034-4877(78)90063-0
12. Duval, C., Künzle, H.P.: Minimal gravitational coupling in the Newtonian theory and the covariant Schrödinger equation. Gen. Relativ. Gravit. **16**(4), 333–347 (1984). https://doi.org/10.1007/BF00762191
13. Ehlers, J.: Über den Newtonschen Grenzwert der Einsteinschen Gravitationstheorie. In: Nitsch, J., Pfarr, J., Stachow, E.W. (eds.) Grundlagenprobleme der modernen Physik: Festschrift für Peter Mittelstaedt zum 50. Geburtstag, pp. 65–84. Bibliographisches Institut, Mannheim, Wien, Zürich (1981). Republished as [14]

14. Ehlers, J.: On the Newtonian limit of Einstein's theory of gravitation. Gen. Relativ. Gravit. **51**(12), 163 (2019). https://doi.org/10.1007/s10714-019-2624-0. Republication of original article [13] as 'Golden Oldie'
15. Friedrichs, K.: Eine invariante Formulierung des Newtonschen Gravitationsgesetzes und des Grenzüberganges vom Einsteinschen zum Newtonschen Gesetz. Math. Ann. **98**, 566–575 (1928). https://doi.org/10.1007/BF01451608
16. Geracie, M., Prabhu, K., Roberts, M.M.: Curved non-relativistic spacetimes, Newtonian gravitation and massive matter. J. Math. Phys. **56**(10), 103505 (2015). https://doi.org/10.1063/1.4932967
17. Geracie, M., Prabhu, K., Roberts, M.M.: Fields and fluids on curved non-relativistic spacetimes. J. High Energy Phys. **2015**(8), 42 (2015). https://doi.org/10.1007/JHEP08(2015)042
18. Geracie, M., Son, D.T., Wu, C., Wu, S.F.: Spacetime symmetries of the quantum Hall effect. Phys. Rev. D **91**(4), 045030 (2015). https://doi.org/10.1103/PhysRevD.91.045030
19. Hansen, D., Hartong, J., Obers, N.A.: Action principle for Newtonian gravity. Phys. Rev. Lett. **122**(6), 061106 (2019). https://doi.org/10.1103/PhysRevLett.122.061106
20. Julia, B., Nicolai, H.: Null-Killing vector dimensional reduction and Galilean geometrodynamics. Nucl. Phys. B **439**(1), 291–323 (1995). https://doi.org/10.1016/0550-3213(94)00584-2
21. Künzle, H.P.: Galilei and Lorentz structures on space-time : Comparison of the corresponding geometry and physics. Ann. Inst. Henri Poincaré **17**(4), 337–362 (1972). http://www.numdam.org/item/AIHPA_1972__17_4_337_0/
22. Künzle, H.P.: Covariant Newtonian limit of Lorentz space-times. Gen. Relativ. Gravit. **7**(5), 445–457 (1976). https://doi.org/10.1007/BF00766139
23. Malament, D.B.: Topics in the Foundations of General Relativity and Newtonian Gravitation Theory. In: Chicago Lectures in Physics. The University of Chicago Press, Chicago (2012). https://press.uchicago.edu/ucp/books/book/chicago/T/bo12893557.html. Available at the author's website under https://www.socsci.uci.edu/~dmalamen/bio/GR.pdf
24. Son, D.T.: Newton-Cartan geometry and the quantum Hall effect (2013). arXiv:1306.0638
25. Trautman, A.: Sur la théorie newtonienne de la gravitation. C. R. Acad. Sci. **257**, 617–620 (1963). https://gallica.bnf.fr/ark:/12148/bpt6k4007z/f639.item
26. Trautman, A.: Foundations and current problems of general relativity. In: Deser, S., Ford, K.W. (eds.) Lectures on General Relativity, vol. 1, pp. 1–248. Prentice-Hall, Englewood Cliffs, N. J. (1965). http://trautman.fuw.edu.pl/publications/Books/21.pdf

Galilei Manifolds

2

Abstract

Here, we will introduce and develop the basic framework for the description of spacetime in Newton–Cartan gravity: so-called Galilei manifolds and Galilei connections on them.

2.1 Foundations of Galilei Manifolds

Definition 2.1 A *Galilei manifold* is an $(n + 1)$-dimensional differentiable manifold M (with $n \geq 1$) together with

(i) a nowhere-vanishing one-form $\tau \in \Omega^1(M)$, called the *clock form*, and
(ii) a symmetric contravariant degree-2 tensor field $h = h^{\mu\nu}\, \partial_\mu \otimes \partial_\nu \in \Gamma(\bigvee^2 TM)$
   which is positive semidefinite of rank n, called the *space metric*, such that
(iii) τ spans the degenerate direction of h, i.e. we have

$$\tau_\mu h^{\mu\nu} = 0, \tag{2.1}$$

A vector $v \in TM$ is *spacelike* if $\tau(v) = 0$, and *timelike* otherwise. It is *future-directed* if $\tau(v) > 0$, and *past-directed* if $\tau(v) < 0$. It is *(future-directed) unit timelike* if $\tau(v) = 1$.

The Galilei manifold has *absolute time* if the clock form is closed, i.e. $d\tau = 0$. ✿

Interpretation 2.2 Points of M are interpreted as spacetime events. τ is interpreted as measuring *time*, i.e. the integral

$$\int_\gamma \tau = \int_{\lambda_i}^{\lambda_f} d\lambda\, \tau(\gamma'(\lambda)) \tag{2.2}$$

© The Author(s), under exclusive license to Springer Nature Switzerland AG 2026
P. K. Schwartz, *Newton–Cartan Gravity*, Lecture Notes in Physics 1044,
https://doi.org/10.1007/978-3-032-03967-5_2

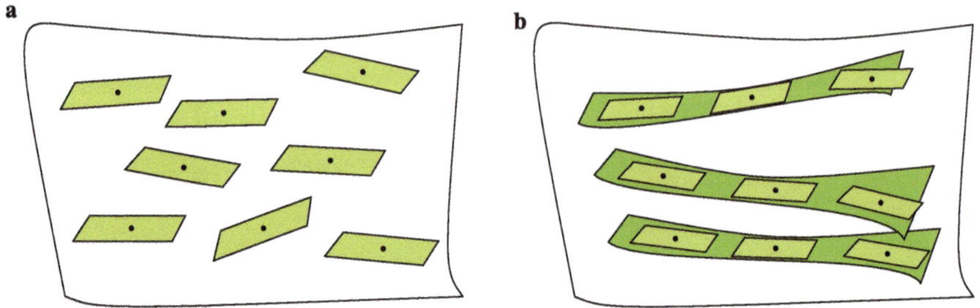

Fig. 2.1 **a** A Galilei manifold with its distribution of spacelike vectors. **b** A Galilei manifold with integrable spacelike distribution, foliated by leaves of space

of τ along a curve $\gamma\colon [\lambda_i, \lambda_f] \to M$ is the time elapsed along γ. Along vectors in $\ker \tau|_p$, 'no time passes', i.e. they point in the direction of events that are ('infinitesimally') *simultaneous* to p. In this way, the distribution $\ker \tau \subset TM$ of spacelike vectors endows M with an infinitesimal version of Newtonian causality; see Fig. 2.1a.

If we assume $\tau \wedge \mathrm{d}\tau = 0$, then by the Frobenius theorem the distribution $\ker \tau$ of spacelike vectors is integrable, i.e. M is foliated by submanifolds whose tangent spaces at each point are the respective spaces of spacelike vectors. Put differently, $\tau \wedge \mathrm{d}\tau = 0$ implies that spacetime is foliated by n-dimensional 'leaves of space'; see Fig. 2.1b. The spacetime then allows for an *absolute notion of simultaneity*, with two events being considered simultaneous if (and only if) they lie on the same leaf of the foliation.

Assuming the even stronger condition $\mathrm{d}\tau = 0$, the time between two events is independent of the curve connecting them (as long as we restrict to simply connected regions in M): we take any two curves γ_1, γ_2 connecting the same two points, and any surface A filling up the boundary curve $\partial A = \gamma_1 \cup \overline{\gamma_2}$ (with the bar over γ_2 denoting reversed orientation), see Fig. 2.2. By Stokes' theorem we then have

$$\int_{\gamma_1} \tau - \int_{\gamma_2} \tau = \int_{\partial A} \tau = \int_A \mathrm{d}\tau = 0, \tag{2.3}$$

i.e. the time along γ_1 is the same as that along γ_2. In this sense, a closed clock form defines an absolute notion of time, which is why we say that a Galilei manifold with closed clock form has absolute time. By the Poincaré lemma, this implies that locally

Fig. 2.2 Two curves connecting the same events

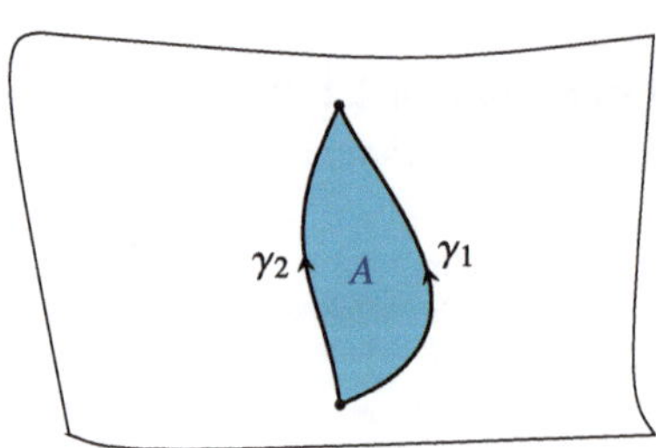

$\tau = dt$ for a function t, such that the spatial leaves from above are hypersurfaces of constant t, and the time between two events is the t difference between them.　　△

Exercise 2.3 (*An example Galilei manifold*) Convince yourself that $M = \mathbb{R}^{n+1} = \{(t, x^1, \ldots, x^n)\}$ with the tensor fields $\tau = dt$, $h = \delta^{ab}\partial_a \otimes \partial_b$ (where $\partial_a = \frac{\partial}{\partial x^a}$ and a, b are summed from 1 to n) is a Galilei manifold. Does it have absolute time?　　△

Remark 2.4

(i) Our definition of Galilei manifolds includes a time orientation: the sign of the clock form τ evaluated on timelike vectors determines if they are future- or past-directed, in a globally consistent way.

One can generalise the definition to allow for Galilei manifolds which are not necessarily time-orientable in the following way: instead of a clock form τ, we work with a symmetric *time metric* $\tilde{\tau} = \tilde{\tau}_{\mu\nu}dx^\mu \otimes dx^\nu \in \Gamma(\bigvee^2 T^*M)$, positive semidefinite of rank 1, such that $\tilde{\tau}$ and h satisfy $\tilde{\tau}_{\mu\nu}h^{\nu\rho} = 0$. The time along a curve γ is then given by

$$\int_\gamma \sqrt{\tilde{\tau}_{\mu\nu}dx^\mu dx^\nu} = \int_{\lambda_i}^{\lambda_f} d\lambda\sqrt{\tilde{\tau}(\gamma'(\lambda), \gamma'(\lambda))}\,, \tag{2.4}$$

spacelike vectors are those with zero '$\tilde{\tau}$-square', and timelike vectors those with positive $\tilde{\tau}$-square. Two timelike vectors v, w point *in the same time direction* if $\tilde{\tau}(v, w) > 0$. We can make a globally consistent choice of time orientation if and only if there is a nowhere-vanishing timelike vector field (which we could then declare to be future-pointing), or equivalently by writing $\tilde{\tau} = \tau \otimes \tau$ in terms of a nowhere-vanishing τ, which would then be the clock form.

For the sake of notational simplicity, we restrict our considerations to time-oriented Galilei manifolds, i.e. work in terms of a clock form. Note, however, that essentially all of our constructions and discussions also work in the case of non-time-orientable Galilei manifolds.

(ii) On a given differentiable manifold M (of dimension ≥ 2), there exist τ and h making it into a Galilei manifold if and only if there exists a nowhere-vanishing vector field. In fact, choose any Riemannian metric g on M. Given a nowhere-vanishing vector field $X \in \Gamma(TM)$, the fields $\tau := g(X, \cdot)$ and $h := g^{-1} - \frac{1}{g(X,X)}X \otimes X$ make M into a Galilei manifold. Conversely, given τ we obtain the nowhere-vanishing vector field $g^{-1}(\tau, \cdot)$.　　△

Construction 2.5 The space metric h defines a positive-definite metric on the distribution of spacelike vectors, i.e. a positive-definite scalar product on spacelike vectors at any point $p \in M$. This works as follows. $h|_p \in \bigvee^2 T_pM \subset T_pM \otimes T_pM$ can be interpreted as a linear map $h|_p : T_p^*M \to T_pM$ (acting as $\alpha \mapsto h^{\mu\nu}|_p\, \alpha_\mu\, \partial_\nu|_p$). By definition this map has a one-dimensional kernel spanned by $\tau|_p$, and therefore induces an isomorphism

$$\widetilde{h}|_p \colon\ T_p^*M \Big/ \operatorname{span}\{\tau|_p\} \to \operatorname{im} h|_p. \tag{2.5}$$

On the other hand, we know that the image of $h|_p$ consists of spacelike vectors, and for dimensional reasons it is the whole space of spacelike vectors, i.e. $\operatorname{im} h|_p = \ker \tau|_p$. Thus we can identify the space of spacelike vectors with the quotient space on the left-hand side of (2.5), on which $h|_p$ induces a positive-definite scalar product.

Concretely this means that for any two spacelike vectors $v,\, w \in \ker \tau|_p \subset T_pM$, there are covectors $\alpha,\, \beta \in T_p^*M$ with $v^\mu = h^{\mu\nu}\alpha_\nu$, $w^\mu = h^{\mu\nu}\beta_\nu$, and the scalar product between v and w is

$$h(\alpha, \beta) = \alpha_\mu w^\mu = \beta_\mu v^\mu \,. \tag{2.6}$$

Here α and β are non-unique up to an addition of multiples of $\tau|_p$, which does not affect the resulting value.

In this way, the 'space metric' really allows to define lengths of (and angles between) spacelike vectors, i.e. to measure *spatial distances* between 'infinitesimally close' simultaneous events.

Note that if the spacelike distribution is integrable (i.e. $\tau \wedge d\tau = 0$), this means that h induces a Riemannian metric on each spatial leaf. $\triangle$

Definition 2.6 For an $(n+1)$-dimensional Galilei manifold (M, τ, h), we will denote the bundle metric induced by h on the n-dimensional spacelike distribution $\ker \tau$ by $^{(n)}h$. ❀

Definition 2.7 A Galilei manifold (M, τ, h) with integrable spacelike distribution is *spatially flat* if each of the spatial leaves foliating M is flat (as a Riemannian manifold with the induced metric $^{(n)}h$). ❀

Construction 2.8 We can invert Construction 2.5: starting only with τ and a positive definite bundle metric $^{(n)}h$ on $\ker \tau$, we can encode the latter in a space metric h in the sense introduced before, as follows. At each point $p \in M$, we have the scalar product $^{(n)}h|_p \in \bigvee^2(\ker \tau|_p)^* \subset (\ker \tau|_p)^* \otimes (\ker \tau|_p)^*$. This induces a linear isomorphism $(^{(n)}h|_p)^{-1} \colon (\ker \tau|_p)^* \overset{\cong}{\to} \ker \tau|_p$. Further, we have a natural surjective linear map $T_p^*M \twoheadrightarrow (\ker \tau|_p)^*$, given by restriction of covectors in T_p^*M to $\ker \tau|_p$. Composing these maps with the inclusion $\ker \tau|_p \hookrightarrow T_pM$, we obtain a linear map

$$h|_p \colon T_p^*M \twoheadrightarrow (\ker \tau|_p)^* \overset{(^{(n)}h|_p)^{-1}}{\underset{\cong}{\longrightarrow}} \ker \tau|_p \hookrightarrow T_pM, \tag{2.7a}$$

i.e. an element $h|_p \in T_pM \otimes T_pM$.

$^{(n)}h|_p$ being symmetric as a bilinear form (on $\ker \tau|_p$) is equivalent to the linear map $^{(n)}h|_p \colon \ker \tau|_p \to (\ker \tau|_p)^*$ being its own dual map.[1] Further, the restriction map $T_p^*M \twoheadrightarrow (\ker \tau|_p)^*$ and the inclusion $\ker \tau|_p \hookrightarrow T_pM$ are duals of each other. Therefore, the linear map in (2.7a) is its own dual. This means that $h|_p$ is symmetric when understood as a bilinear form on T_p^*M.

Less abstractly, this argument may be formulated as follows: the bilinear form $h|_p$ on T_p^*M is defined by

$$h|_p(\alpha, \beta) = \big(^{(n)}h|_p\big)^{-1}\big(\alpha|_{\ker \tau|_p}, \beta|_{\ker \tau|_p}\big) \tag{2.7b}$$

in terms of the form $\big(^{(n)}h|_p\big)^{-1}$ on $(\ker \tau|_p)^*$, which is the inverse of the non-degenerate $^{(n)}h|_p$. Since the latter is symmetric, $h|_p$ as defined by (2.7b) is symmetric.

By construction, the kernel of the linear map in (2.7a) is precisely $\mathrm{span}\{\tau|_p\}$; i.e. the symmetric bilinear form $h|_p$ is degenerate precisely on $\mathrm{span}\{\tau|_p\}$. Finally, for any $\alpha \in T_p^*M \setminus \mathrm{span}\{\tau|_p\}$, writing $\tilde{\alpha} := \alpha|_{\ker \tau|_p} \in (\ker \tau|_p)^* \setminus \{0\}$, by (2.7b) we have $h|_p(\alpha, \alpha) = \big(^{(n)}h|_p\big)^{-1}(\tilde{\alpha}, \tilde{\alpha}) > 0$, since $^{(n)}h|_p$ is positive definite. Hence, $h|_p$ is positive semidefinite. $\triangle$

Constructions 2.5 and 2.8 show that equivalently we could have defined Galilei manifolds in terms of a clock form τ and a positive definite bundle metric $^{(n)}h$ on $\ker \tau$. Since $^{(n)}h$ encodes spatial lengths, from an 'operational' point of view this would be a more natural definition. However, working with the spacetime tensor field h (instead of $^{(n)}h$, which is only defined on a subbundle) is mathematically easier, which is why we have chosen the definition in terms of τ and h.

Construction 2.9 Let (M, τ, h) be a Galilei manifold and v a unit timelike vector field on it,[2] i.e. a vector field $v \in \Gamma(TM)$ with $\tau(v) = 1$. With respect to v, at each point $p \in M$ the tangent and cotangent spaces decompose as

$$T_pM = \mathrm{span}\{v|_p\} \oplus \ker \tau|_p \,, \tag{2.8a}$$

$$T_p^*M = \mathrm{span}\{\tau|_p\} \oplus \ker v|_p \,. \tag{2.8b}$$

Hence, there is a unique projection operator $T_pM \to T_pM$ that projects onto the space $\ker \tau|_p$ of spacelike vectors according to the decomposition (2.8a), i.e. with kernel $\mathrm{span}\{v|_p\}$.

Using the decompositions (2.8), we can also define a kind of 'inverse' to h: the map $h|_p \colon T_p^*M \to T_pM$ has kernel $\mathrm{span}\{\tau|_p\}$ and rank n, so it maps $\ker v|_p$ isomorphically onto $\ker \tau|_p$. Therefore, each (spacelike) vector in $\ker \tau|_p$ has a unique

---

[1] Here, the dual of a linear map $f \colon V \to W$ is $f^* \colon W^* \to V^*$ defined by $f^*(\alpha) = \alpha \circ f$ for $\alpha \in W^*$. For V finite-dimensional, we have $V^{**} = V$ in a natural way, so the dual of a map $V^* \to V$ is again a map $V^* \to V^{**} = V$. Hence, for a map $V^* \to V$ it makes sense to be its own dual.

[2] Some of the following discussion is in great parallel to that in my (the author's) article [2].

preimage in $\ker v|_p$ under $h|_p$ (compare Construction 2.5). Hence, there is a unique linear map $T_pM \to T_p^*M$ that vanishes on $\mathrm{span}\{v|_p\}$ and on $\ker \tau|_p$ maps each vector to its preimage under $h|_p$ in $\ker v|_p$. $\qquad\qquad\qquad\qquad\qquad\triangle$

Definition 2.10 Let (M, τ, h) be a Galilei manifold and v a unit timelike vector field on it.

(i) The *spatial projector along* v is the unique endomorphism P of TM that is a projection operator, projects onto $\ker \tau$, and whose kernel is spanned by v. Explicitly, it is given by

$$P = \mathrm{id} - v \otimes \tau, \tag{2.9a}$$

or, in components,

$$P^\mu_\nu := \delta^\mu_\nu - v^\mu \tau_\nu . \tag{2.9b}$$

Note that even though it depends on the choice of v, we will not acknowledge that in the notation, simply to avoid cluttering equations.

(ii) The *covariant space metric with respect to* v is the symmetric covariant 2-tensor field $\underset{v}{h} = h_{\mu\nu}\, dx^\mu \otimes dx^\nu \in \Gamma(\bigvee^2 T^*M)$ which, understood as a vector bundle homomorphism $TM \to T^*M$, vanishes on $\mathrm{span}\{v\}$ and on $\ker \tau$ maps each vector to its preimage under $h\colon T^*M \to M$ in $\ker v$. In index notation, this means that is has to satisfy

$$h_{\mu\nu}v^\nu = 0, \;\; h_{\mu\nu}h^{\nu\rho} = P^\rho_\mu . \tag{2.10}$$

Note that in index notation, we leave out the subscript v in order to avoid confusion with an index. An h with indices 'downstairs' will always mean the covariant space metric with respect to that unit timelike vector field v which is clear from context. $\qquad\qquad\qquad\qquad\qquad\qquad\qquad\qquad\maltese$

Note that Construction 2.9 also implies that the metric $^{(n)}h$ induced by h on spacelike vectors may be expressed as $^{(n)}h(w, \tilde{w}) = \underset{v}{h}(w, \tilde{w})$ for any two spacelike vectors $w, \tilde{w} \in \ker \tau|_p$, i.e. we have $^{(n)}h = \underset{v}{h}|_{\ker \tau}$.

Definition 2.11 The decomposition of vector fields and one-forms on M according to (2.8) we call the decomposition into *timelike* and *spacelike parts with respect to* v. Explicitly, for a vector field X and a one-form α, these decompositions are

$$X^\mu = \delta^\mu_\nu X^\nu = v^\mu \tau_\nu X^\nu + P^\mu_\nu X^\nu, \tag{2.11a}$$

$$\alpha_\mu = \delta^\nu_\mu \alpha_\nu = \tau_\mu v^\nu \alpha_\nu + P^\nu_\mu \alpha_\nu , \tag{2.11b}$$

where the first term is the timelike and the second term the spacelike part with respect to v. We may also apply this decomposition to higher-degree tensors/tensor fields. $\maltese$

Notation 2.12 When employing index notation for tensor fields on a Galilei manifold, we will raise indices using h. For example, if we have some tensor field with components $X^\mu{}_{\nu\rho}{}^\sigma{}_\kappa$, we use the notation

$$X^{\mu\nu}{}_{\rho\ \kappa}^{\ \ \sigma} := h^{\nu\lambda} X^\mu{}_{\lambda\rho}{}^\sigma{}_\kappa \ . \tag{2.12a}$$

Furthermore, when a unit timelike vector field v is chosen, we will lower indices using $\underset{v}{h}$. For example, for the tensor field from above we will write

$$X_{\mu\nu\rho}{}^\sigma{}_\kappa := \underset{v}{h}_{\mu\lambda} X^\lambda{}_{\nu\rho}{}^\sigma{}_\kappa \ . \tag{2.12b}$$

Note that since h and $\underset{v}{h}$ are degenerate, the operations of raising and lowering indices are not invertible, i.e. differently to the case in (pseudo-)Riemannian geometry we lose information when doing so. First lowering an index (with $\underset{v}{h}$) and then raising it again (with h), or vice versa, corresponds to contracting it with the spatial projector P along v. Therefore, we have to keep in mind the original position of indices in order not to forget these projections (except for spacelike indices, for which lowering and raising are actual inverses of each other). ✿

Finally, we want to introduce bases of the tangent space that are adapted to the structure of a Galilei manifold.

Definition 2.13 Let (M, τ, h) be a Galilei manifold. A *Galilei basis* at a point $p \in M$ is an (ordered) basis $(e_A) = (e_t, e_a)$, $a = 1, \ldots, n$, of the tangent space $T_p M$ such that

(i) for the dual basis $(e^A) = (e^t, e^a)$ of $T_p^* M$, we have $e^t = \tau|_p$, and
(ii) $h|_p = \delta^{ab} e_a \otimes e_b$, i.e. in components $h^{\mu\nu}|_p = \delta^{ab} e_a^\mu e_b^\nu$. ✿

Notation 2.14 When labelling the elements of a Galilei basis or the corresponding dual basis, we use t as a 'temporal' index, labelling the first element, and lower-case Latin letters from the beginning of the alphabet as 'spatial' indices running from 1 to n, labelling the rest. Capital Latin indices run through the full range $\{t, 1, \ldots, n\}$. ✿

Remark 2.15 Note that *either* of the two conditions from Definition 2.13 implies that the vectors e_a are spacelike: using the definition of the dual basis, from the first condition we obtain $\tau(e_a) = e^t(e_a) = 0$. By definition of a Galilei manifold, the second condition gives $0 = h|_p(\tau, \cdot) = \delta^{ab}\tau(e_a)e_b$, which also implies $\tau(e_a) = 0$. △

Proposition 2.16 *In the definition of a Galilei basis, condition (i) may be replaced by the following:*

(i') e_t is a future-directed unit timelike vector, i.e. $\tau(e_t) = 1$.

Similarly, as long as (i) is kept, condition (ii) may be replaced by the following:

(ii')   *The 'spatial' elements e^a of the dual basis $(e^A) = (e^t, e^a)$ are orthonormal with respect to h, i.e. we have $h(e^a, e^b) = \delta^{ab}$.*

Finally, condition (ii) may be replaced—in combination with either (i) or (i')—by the following:

(ii")   *The vectors e_a are spacelike, i.e. $\tau(e_a) = 0$, and they are orthonormal with respect to the induced metric $^{(n)}h|_p$ on $\ker \tau|_p$, i.e. $^{(n)}h(e_a, e_b) = \delta_{ab}$.*

Proof Given a basis satisfying (i), by the definition of the dual basis we have $\tau(e_t) = e^t(e_t) = 1$, i.e. (i'). Conversely, given (i') and (ii), we have $\tau(e_t) = 1$ and (as we have seen above) $\tau(e_a) = 0$, which implies $\tau|_p = e^t$, i.e. (i).

Now for (ii'). Given a basis satisfying (ii), the definition of the dual basis gives us $h(e^a, e^b) = \delta^{cd}(e_c \otimes e_d)(e^a, e^b) = \delta^{cd}\delta^a_c\delta^b_d = \delta^{ab}$, which is (ii'). Conversely, given (i) and (ii'), we have $0 = h(\tau|_p, \cdot) = h(e^t, \cdot)$, such that h is given by $h|_p = h(e^a, e^b)e_a \otimes e_b = \delta^{ab}e_a \otimes e_b$, i.e. (ii).

Now we consider (ii"). We are going to show that (ii) implies (ii"), that (ii") implies (ii'), and that (i') together with (ii") implies (i). Together with the previous parts this shows that being a Galilei basis is equivalent to the combination of (i) and (ii") as well as to the combination of (i') and (ii").

First, given a basis satisfying (ii), we have seen above that this implies that the e_a are spacelike. We also obtain $h(\delta_{ab}e^b, \cdot) = \delta^{cd}(e_c \otimes e_d)(\delta_{ab}e^b, \cdot) = \delta^{cd}\delta_{ab}\delta^b_c e_d = e_a$, i.e. $e_a = h(\delta_{ac}e^c, \cdot)$. By the definition of $^{(n)}h$ (Construction 2.5), this implies $^{(n)}h(e_a, e_b) = \delta_{ac}e^c(e_b) = \delta_{ab}$, showing the second part of (ii").

Next, assume our basis satisfies (ii"). For $c \in \{1, \ldots, n\}$, consider the spacelike vector $w_c := \delta_{ca}h(e^a, \cdot)$. The definition of $^{(n)}h$ implies that $^{(n)}h(w_c, e_b) = \delta_{ca}e^a(e_b) = \delta_{cb}$. On the other hand, by assumption we have $^{(n)}h(e_c, e_b) = \delta_{cb}$, so $^{(n)}h$ being non-degenerate implies $w_c = e_c$. Thus, we obtain $\delta_{ca}h(e^a, e^b) = e^b(w_c) = e^b(e_c) = \delta^b_c$, or equivalently $h(e^a, e^b) = \delta^{ab}$, showing (ii').

Finally, assuming (i') and (ii"), we have $\tau(e_t) = 1$ and $\tau(e_a) = 0$, implying $\tau|_p = e^t$, i.e. (i). $\square$

Note that (i') and (ii') together are *weaker* than the original conditions, i.e. they do *not* characterise a Galilei basis:

Exercise 2.17 Let (M, τ, h) be a Galilei manifold and $(\tilde{e}_t, \tilde{e}_a)$ a Galilei basis at $p \in M$. Show that the new basis (e_t, e_a) defined by $e_t := \tilde{e}_t - \frac{1}{2}\tilde{e}_1$, $e_1 := \tilde{e}_t + \frac{1}{2}\tilde{e}_1$, and $e_a := \tilde{e}_a$ for $a > 1$ satisfies (i') and (ii'), but is not a Galilei basis.

Hint: You can show (i') by direct computation, using the definition of the dual basis. For (ii'), express the $\tilde{e}_a$ in terms of the new basis and insert this into $h|_p = \delta^{ab}\tilde{e}_a \otimes \tilde{e}_b$; using this result, (ii') follows by computation. Finally use that if the new basis were a Galilei basis, all of the e_a would have to be spacelike. $\triangle$

The different characterisations of Galilei bases we saw above have slightly different 'flavours'. The original definition by (i) and (ii) characterises Galilei bases as bases with respect to which the tensors $\tau|_p$ and $h|_p$ may be expressed easily, so it has the most direct connection to our definition of Galilei manifolds in terms of τ and h. In contrast to this 'formal' characterisation, that given by (i') and (ii") is more 'operational': it speaks only about the basis vectors themselves, not the dual basis, and demands them to be a unit timelike vector on the one hand and an orthonormal basis of spacelike vectors on the other hand—statements about, essentially, (idealised) direct physical measurements. The characterisation by (i) and (ii') is, in contrast, given purely in terms of the dual basis, and it is again more 'formal' in nature: the first dual basis vector has to be $\tau|_p$, and the whole dual basis a Sylvester basis for $h|_p$. (Note that this also shows most directly the existence of Galilei bases.) The remaining two characterisations by (i') and (ii), and by (i) and (ii") are somewhat unnatural combinations of aspects of the previous ones.

Proposition 2.18 *Let (M, τ, h) be a Galilei manifold and (e_A) a Galilei basis at $p \in M$.*

(i) The spatial projector[3] along e_t can be expressed as

$$P = e_a \otimes e^a \text{ or, in components, } P^\mu_\nu = e^\mu_a e^a_\nu . \tag{2.13}$$

(ii) The covariant space metric with respect to e_t can be expressed as

$$\underset{e_t}{h} = \delta_{ab} e^a \otimes e^b \text{ or, in components, } h_{\mu\nu} = \delta_{ab} e^a_\mu e^b_\nu . \tag{2.14}$$

Proof (i) The definition of the dual basis can be expressed in the form $\mathrm{id}_{T_p M} = e_A \otimes e^A = e_t \otimes e^t + e_a \otimes e^a$. By the definition of the projector, the result follows.

(ii) Contracting $\delta_{ab} e^a_\mu e^b_\nu$ with e^μ_t and with $h^{\nu\rho}$, we obtain

$$\delta_{ab} e^a_\mu e^b_\nu e^\mu_t = 0 \tag{2.15}$$

and

$$\delta_{ab} e^a_\mu e^b_\nu h^{\nu\rho} = \delta_{ab} e^a_\mu e^b_\nu \delta^{cd} e^\nu_c e^\rho_d = \delta_{ab} e^a_\mu \delta^b_c \delta^{cd} e^\rho_d = e^a_\mu e^\rho_a = P^\rho_\mu , \tag{2.16}$$

i.e. we have shown the defining properties of the covariant space metric. $\qquad\square$

---

[3] We have only defined this projector for unit timelike vector *fields*, but of course the definition also works just at a single point. The same goes for the covariant space metric.

2.2   Galilei Connections

Similarly to the well-known case of (pseudo-)Riemannian manifolds, we will now consider connections compatible with the structure of a Galilei manifold.

Definition 2.19 A *Galilei connection* on a Galilei manifold (M, τ, h) is a linear connection ∇ satisfying

$$\nabla \tau = 0, \ \nabla h = 0. \tag{2.17}$$

❀

Proposition 2.20 *Let ∇ be a Galilei connection on (M, τ, h).*

(i) The 'timelike torsion' of ∇ is $\mathrm{d}\tau$, i.e. we have

$$\tau(T(X, Y)) = \mathrm{d}\tau(X, Y) \tag{2.18a}$$

for any vectors (or vector fields) X, Y, or

$$\tau_\rho T^\rho{}_{\mu\nu} = (\mathrm{d}\tau)_{\mu\nu} \tag{2.18b}$$

expressed in index notation.
(ii) The curvature tensor of ∇ satisfies $\tau_\mu R^\mu{}_{\nu\rho\sigma} = 0$ and $R^{\mu\nu}{}_{\rho\sigma} = -R^{\nu\mu}{}_{\rho\sigma}$.

Proof (i) Using compatibility of ∇ with τ, for any vector fields X, Y we obtain

$$\begin{aligned}
\tau(T(X, Y)) &= \tau(\nabla_X Y) - \tau(\nabla_Y X) - \tau([X, Y]) \\
&= \nabla_X(\tau(Y)) - \nabla_Y(\tau(X)) - \tau([X, Y]) \\
&= X(\tau(Y)) - Y(\tau(X)) - \tau([X, Y]) \\
&= \mathrm{d}\tau(X, Y).
\end{aligned} \tag{2.19}$$

(ii) Again using compatibility of ∇ with τ, for any vector fields X, Y, Z we have

$$\begin{aligned}
\tau(R(X, Y)Z) &= \tau(\nabla_X \nabla_Y Z) - \tau(\nabla_Y \nabla_Y Z) - \tau(\nabla_{[X,Y]} Z) \\
&= X(\tau(\nabla_Y Z)) - Y(\tau(\nabla_X Z)) - [X, Y](\tau(Z)) \\
&= X(Y(\tau(Z))) - Y(X(\tau(Z))) - [X, Y](\tau(Z)) \\
&= 0.
\end{aligned} \tag{2.20}$$

Expressed in index notation, this gives the first identity.

The compatibility of ∇ with h leads to the antisymmetry of the curvature tensor in the first two indices when raised, the details of which we leave to Exercise 2.21 (b). $\qquad\square$

In particular, we see that if $d\tau \neq 0$, any Galilei connection necessarily has torsion. Put differently, torsion-free Galilei connections can exist only on Galilei manifolds with absolute time.

Exercise 2.21 (a) Give a proof of Proposition 2.20 (i) using index notation.

Hint: Use that $(d\tau)_{\mu\nu} = 2\partial_{[\mu}\tau_{\nu]}$ *and* $T^{\rho}{}_{\mu\nu} = 2\Gamma^{\rho}{}_{[\mu\nu]}$.
(b) Prove the second part of Proposition 2.20 (ii), i.e. show that the curvature tensor of any Galilei connection satisfies $R^{\mu\nu}{}_{\rho\sigma} = -R^{\nu\mu}{}_{\rho\sigma}$.

Hint: In index notation, the curvature tensor of any connection satisfies the identity

$$R^{\mu}{}_{\nu\rho\sigma} X^{\nu} = 2\nabla_{[\rho}\nabla_{\sigma]}X^{\mu} + T^{\nu}{}_{\rho\sigma}\,\nabla_{\nu}X^{\mu} \tag{2.21}$$

for any vector field X ('Ricci identity'). Use this to evaluate the expression $2\nabla_{[\rho}\nabla_{\sigma]}(X^{\mu}Y^{\nu})$ for two arbitrary vector fields X, Y. Applying your result to $h^{\mu\nu}$ (which can be written as a sum of tensor products of two vector fields) and using $\nabla h = 0$, you should obtain the desired identity. $\triangle$

Construction 2.22 We consider the spacelike distribution $\ker\tau$. We claim that due to $\nabla\tau = 0$, any Galilei connection ∇ induces a natural connection $\overset{(n)}{\nabla}$ on $\ker\tau$ by restriction.

Indeed, given any vector $v \in T_pM$ and spacelike vector field X, the covariant derivative $\nabla_v X \in T_pM$ is again spacelike: we have $\tau|_p(\nabla_v X) = v(\tau(X)) - (\nabla_v\tau)(X|_p) = 0$ since X is spacelike.

Additionally, due to $\nabla h = 0$, the induced connection $\overset{(n)}{\nabla}$ is compatible with the spatial bundle metric $^{(n)}h$ induced by h:

Let X be any vector field and Y, Z be spacelike vector fields. There exists a one-form α such that

$$Y = h(\alpha, \cdot). \tag{2.22}$$

The induced metric evaluated on Y is defined in terms of α by $^{(n)}h(Y, Z) = \alpha(Z)$. For an arbitrary one-form κ, (2.22) means that $\kappa(Y) = h(\alpha, \kappa)$. Therefore, we can conclude

$$\kappa\left(\overset{(n)}{\nabla}_X Y\right) = \kappa(\nabla_X Y)$$

$$= X(\kappa(Y)) - (\nabla_X\kappa)(Y)$$

$$= X(h(\alpha, \kappa)) - h(\alpha, \nabla_X\kappa)$$

$$= h(\nabla_X\alpha, \kappa), \tag{2.23}$$

where in the last step we used $\nabla h = 0$. We thus have shown that

$$\overset{(n)}{\nabla}_X Y = h(\nabla_X\alpha, \cdot). \tag{2.24}$$

By definition of $^{(n)}h$, this implies

$$^{(n)}h\left(\overset{(n)}{\nabla}_X Y, Z\right) = (\nabla_X \alpha)(Z). \tag{2.25}$$

Hence, we obtain

$$\begin{aligned}
X\left(^{(n)}h(Y, Z)\right) &= X(\alpha(Z)) \\
&= (\nabla_X \alpha)(Z) + \alpha(\nabla_X Z) \\
&= (\nabla_X \alpha)(Z) + \alpha\left(\overset{(n)}{\nabla}_X Z\right) \\
&= {}^{(n)}h\left(\overset{(n)}{\nabla}_X Y, Z\right) + {}^{(n)}h\left(Y, \overset{(n)}{\nabla}_X Z\right),
\end{aligned} \tag{2.26}$$

which is compatibility of $\overset{(n)}{\nabla}$ with $^{(n)}h$. $\qquad\qquad\qquad\qquad\qquad\qquad\triangle$

Corollary 2.23 *On a Galilei manifold (M, τ, h) with integrable spacelike distribution, i.e. satisfying $\tau \wedge \mathrm{d}\tau = 0$, any Galilei connection ∇ induces a metric connection $\overset{(n)}{\nabla}$ on each of the spatial leaves (which are Riemannian manifolds). In the case of a torsion-free Galilei connection on a Galilei manifold with absolute time, the induced connections $\overset{(n)}{\nabla}$ are the Levi-Civita connections of the spatial leaves.* $\qquad\square$

Construction 2.24 According to Construction 2.22 and Corollary 2.23, a torsion-free Galilei connection ∇ on a Galilei manifold (M, τ, h) with absolute time restricts on each spatial leaf to the Levi-Civita connection $\overset{(n)}{\nabla}$ of the induced Riemannian metric $^{(n)}h$.

Therefore, we can restrict the curvature tensor of ∇ to each of the spatial leaves, and when doing so, we obtain the curvature tensor $^{(n)}R$ of the spatial leaf as a Riemannian manifold: for any spatial leaf Σ and vector fields $X, Y, Z \in \Gamma(T\Sigma)$, on Σ we have

$$\begin{aligned}
R(X, Y)Z &= \nabla_X \nabla_Y Z - \nabla_Y \nabla_X Z - \nabla_{[X,Y]}Z \\
&= \nabla_X \overset{(n)}{\nabla}_Y Z - \nabla_Y \overset{(n)}{\nabla}_X Z - \nabla_{[X,Y]}Z \\
&= \overset{(n)}{\nabla}_X \overset{(n)}{\nabla}_Y Z - \overset{(n)}{\nabla}_Y \overset{(n)}{\nabla}_X Z - \overset{(n)}{\nabla}_{[X,Y]}Z \\
&= {}^{(n)}R(X, Y)Z
\end{aligned} \tag{2.27}$$

(where of course to be able to apply ∇, we have to extend X, Y, Z to vector fields in a neighbourhood of Σ in M in some way, the choice of which does not affect the results when restricted to Σ).

Note that this implies that (M, τ, h) being spatially flat—i.e. each spatial leaf being flat as a Riemannian manifold—is equivalent to vanishing of the purely spatial part of the curvature tensor of any torsion-free Galilei connection, i.e. to the curvature satisfying

$$R^{\mu\nu\rho\sigma} = 0. \tag{2.28}$$

Furthermore, when restricting the Ricci tensor Ric of ∇ to any spatial leaf, we obtain the Ricci tensor $^{(n)}\mathrm{Ric}$ of the leaf as a Riemannian manifold: for any two spacelike vectors $v, w \in \ker \tau|_p$, we have[4]

$$
\begin{aligned}
^{(n)}\mathrm{Ric}(v, w) &= \mathrm{tr}_{\ker \tau|_p}\left(^{(n)}R(\cdot, w)v\right) \\
&= \mathrm{tr}_{\ker \tau|_p}(R(\cdot, w)v) \\
(\tau_\mu R^\mu{}_{\nu\rho\sigma} = 0) \quad &= \mathrm{tr}_{T_pM}(R(\cdot, w)v) \\
&= \mathrm{Ric}(v, w).
\end{aligned}
\tag{2.29}
$$

△

Remark 2.25 In Interpretation 2.2 we discussed different integrability conditions on τ and interpreted them in terms of the existence of absolute notions of simultaneity or time on our Galilei manifold. For these interpretations, we can now give an additional heuristic 'explanation' using the relationship between $\mathrm{d}\tau$ and the torsion of a Galilei connection (Proposition 2.20 (i)).

Intuitively speaking, the torsion T of a linear connection ∇ measures how much infinitesimal parallelograms built by parallel transport with ∇ fail to close, see Fig. 2.3: given two vectors $v, w \in T_pM$, we can interpret them as paths to events that are 'infinitesimally close' to p. We parallely transport w along the path v, obtaining a vector $\mathrm{Par}_v(w)$ living at the event to which v points. This vector again points towards an event. Reversing the roles of v and w, we first obtain a vector $\mathrm{Par}_w(v)$ by parallely transporting v along w, and can then consider the event towards which this vector points. The two events at the 'ends' of $\mathrm{Par}_v(w)$ and $\mathrm{Par}_w(v)$ are then infinitesimally close to each other, but they don't necessarily coincide. The vector pointing from the first to the second is the torsion $T(v, w)$.[5]

---

[4] In index notation in a Galilei basis, the calculation takes the form $^{(n)}R_{ab}v^a w^b = {}^{(n)}R^c{}_{acd}v^a w^b = R^c{}_{acb}v^a w^b = R^\mu{}_{a\mu b}v^a w^b = R_{\mu\nu}v^\mu w^\nu$.

[5] This idea may be made rigorous as follows. Given two vectors $v, w \in T_pM$, let $t \mapsto \gamma(t)$ be any curve through p with derivative v there (i.e. $\gamma(0) = p$, $\dot\gamma(0) = v$), and let $W(t)$ be the parallel transport of w along γ for time t (i.e. $W(t) \in T_{\gamma(t)}M$, $W(0) = w$, $\nabla_{\dot\gamma}W(t) = 0$). Further, for each t, let $s \mapsto \eta_t(s)$ be a curve through $\gamma(t)$ with derivative $W(t)$ there. Analogously, let $s \mapsto \tilde\gamma(s)$ be a curve through p with derivative w there, let $V(s)$ be the parallel transport of v along $\tilde\gamma$ for time s, and for each s let $t \mapsto \tilde\eta_s(t)$ be a curve through $\tilde\gamma(s)$ with derivative $V(s)$ there. For t, s small enough, we can reach $\tilde\eta_s(t)$ by a unique geodesic starting at $\eta_t(s)$, whose initial tangent vector is $\exp^{-1}_{\eta_t(s)}(\tilde\eta_s(t)) \in T_{\eta_t(s)}M$. This vector we can parallely transport back to p, first along η_t in reverse direction and then along γ in reverse direction. The resulting vector in T_pM we denote by $C_{v,w}(t, s)$. Of the thus defined map $(t, s) \mapsto C_{v,w}(t, s) \in T_pM$, one can show that

$$
C_{v,w}(0, 0) = 0, \quad \left.\frac{\partial C_{v,w}(t, s)}{\partial t}\right|_{t=s=0} = 0, \quad \left.\frac{\partial C_{v,w}(t, s)}{\partial s}\right|_{t=s=0} = 0, \quad \left.\frac{\partial^2 C_{v,w}(t, s)}{\partial s \partial t}\right|_{t=s=0} = T(v, w),
$$

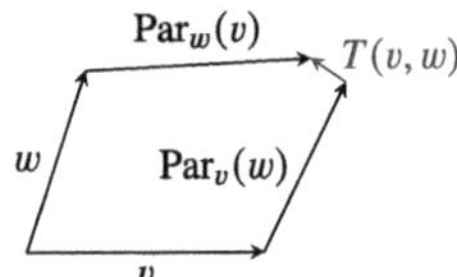

Fig. 2.3 Torsion as closure failure of infinitesimal parallelograms

Let now (M, τ, h) be a Galilei manifold and ∇ a Galilei connection on it. By Proposition 2.20 (i), the torsion of ∇ satisfies $d\tau = \tau(T(\cdot, \cdot))$.

(i) First we consider the condition that the spacelike distribution $\ker \tau$ be integrable, $\tau \wedge d\tau = 0$. This is equivalent to $d\tau$ being of the form $d\tau = \tau \wedge \alpha$ for some one-form α, which is equivalent to $d\tau$ vanishing if both of its arguments are spacelike. Equivalently, $\tau(T(v, w)) = 0$ if both v and w are spacelike, i.e. the torsion evaluated on two spacelike vectors is spacelike. So this condition is equivalent to the statement that *spacelike infinitesimal parallelograms* (i.e. those constructed from spacelike vectors) *close temporally* (the failure to close can only be spacelike).

If this is satisfied, we can consistently define a notion of simultaneity by following spacelike vectors: following the two 'branches' of any spacelike parallelogram, we end up at two infinitesimally close events with a spacelike separation vector. Hence it is consistent to declare these events to be simultaneous to that at the 'origin' of the parallelogram. Of course, this is a very non-rigorous argument; however it offers a somewhat intuitive picture for the connection between the condition $\tau \wedge d\tau = 0$ and the existence of an absolute notion of simultaneity.

(ii) Now we consider the condition that τ be closed, $d\tau = 0$. This is equivalent to $\tau(T(v, w)) = 0$ for any two vectors v, w, i.e. to all torsion being spacelike. So $d\tau = 0$ is equivalent to the statement that *all infinitesimal parallelograms close temporally*.

We can use this to consistently define the temporal distance between (infinitesimally close) spatial leaves, see Fig. 2.4: we start with an event $p \in M$ and a timelike vector $v \in T_p M$. Following v, we end up at an event on a different spatial leaf than the one we started on (but 'infinitesimally close' to it), and $\tau(v)$ is the temporal distance between the leaves measured at these events. In addition, we now take a spacelike vector $w \in T_p M$, and consider the 'infinitesimal parallelogram' formed by v and w. We now follow $\text{Par}_v(w)$ from the 'endpoint' of v, and w from p on the original leaf, ending up at two new events on the two leaves. We want to determine the temporal distance between the two leaves measured at

---

i.e.
$$C_{v,w}(t, s) = st\, T(v, w) + \text{(terms at least quadratic in } t \text{ or } s):$$
the leading term of the 'connecting vector' is given by the torsion.

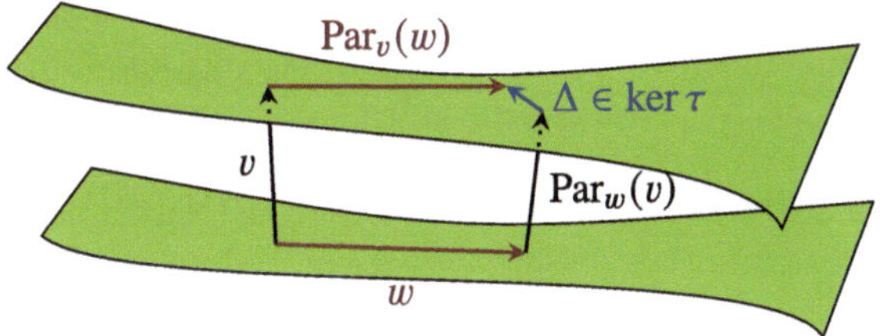

Fig. 2.4 Definition of temporal distance between spatial leaves by parallel transport in the absence of timelike torsion

these two new events. Since the parallelogram's 'closure failure vector' (labelled Δ in Fig. 5) is spacelike, it is consistent to assign as temporal distance between these two events the temporal length $\tau(\mathrm{Par}_w(v))$ of $\mathrm{Par}_w(v)$. Since ∇ is compatible with τ, parallel transport conserves temporal length of vectors, such that $\tau(\mathrm{Par}_w(v)) = \tau(v)$: the temporal distance between the two spatial leaves measured at the displaced events is the same as that measured at the original events. This shows (heuristically) how the condition $\mathrm{d}\tau = 0$, expressed in terms of torsion, allows to define a globally consistent temporal distance between spatial leaves, thus giving rise to an absolute notion of time. $\triangle$

2.3 Classification of Galilei Connections

We now want to classify Galilei connections on a given Galilei manifold (M, τ, h).[6]

Proposition 2.26 *The set of Galilei connections on a Galilei manifold (M, τ, h) is an affine space modelled on the vector space*

$$\left\{ S \in \Gamma(TM \otimes T^*M \otimes T^*M) \mid S^{(\rho}{}_\mu{}^{\nu)} = 0, \ \tau_\rho S^\rho{}_{\mu\nu} = 0 \right\}. \tag{2.30}$$

More precisely, let ∇ be a Galilei connection on (M, τ, h) and $\widetilde{\nabla}$ an arbitrary connection on M, and consider their difference tensor field $S \in \Gamma(TM \otimes T^*M \otimes T^*M)$ defined by $S^\rho{}_{\mu\nu} := \widetilde{\Gamma}^\rho{}_{\mu\nu} - \Gamma^\rho{}_{\mu\nu}$. Then $\widetilde{\nabla}$ is a Galilei connection if and only if S lies in (2.30).

Proof Exercise 2.27. $\square$

Exercise 2.27 Prove Proposition 2.26. $\triangle$

---

[6] Relevant parts of the discussion in this section are in great parallel to my (the author's) article [2], which treats a generalisation.

Definition 2.28 Consider a Galilei connection ∇ on a Galilei manifold (M, τ, h) and a unit timelike vector field v on it. The *Newton–Coriolis form*[7] *of ∇ with respect to v* is the two-form Ω defined by

$$\Omega_{\mu\nu} := 2\big(\nabla_{[\mu}v^{\rho}\big)h_{\nu]\rho} \, . \tag{2.31}$$

❀

Note that the index ρ of $\nabla_{\mu}v^{\rho}$ is spacelike (due to $\nabla\tau = 0$ and $\nabla_{\mu}1 = 0$), such that lowering it with $h_{\nu\rho}$ keeps all information. The antisymmetrisation however leads to Ω encoding less information than ∇v.

In component-free notation, Ω can be characterised by satisfying

$$\Omega(w, u) = {}^{(n)}h(\nabla_w v, u) - {}^{(n)}h(\nabla_u v, w), \tag{2.32a}$$

$$\Omega(v, w) = {}^{(n)}h(\nabla_v v, w) \tag{2.32b}$$

for all spacelike vectors $w, u \in \ker \tau|_p$ at all points $p \in M$. Note that due to antisymmetry of Ω, it is in fact determined by knowing its values on two spacelike vectors as well as on v and one spacelike vector: general arguments of Ω may be decomposed into spacelike and timelike parts with respect to v, and applied to two such decompositions we have

$$\Omega(\lambda v + w, \mu v + u) = \lambda\Omega(v, u) - \mu\Omega(v, w) + \Omega(w, u). \tag{2.33}$$

Using the Newton–Coriolis form, we can now state the classification theorem for Galilei connections:

Theorem 2.29 *(Classification of Galilei connections)   Let (M, τ, h) be a Galilei manifold and v a unit timelike vector field on it.*

(i) *Let ∇ be a Galilei connection on (M, τ, h), T its torsion, and Ω its Newton–Coriolis form with respect to v. Then the torsion satisfies the identity*

$$\tau_{\rho}T^{\rho}{}_{\mu\nu} = (\mathrm{d}\tau)_{\mu\nu} \, , \tag{2.34}$$

and the connection coefficients of ∇ take the form

$$\Gamma^{\rho}_{\mu\nu} = v^{\rho}\partial_{\mu}\tau_{\nu} + \frac{1}{2}h^{\rho\sigma}(2\partial_{(\mu}h_{\nu)\sigma} - \partial_{\sigma}h_{\mu\nu}) + \frac{1}{2}P^{\rho}_{\lambda}T^{\lambda}{}_{\mu\nu} - T_{(\mu\nu)}{}^{\rho} + \tau_{(\mu}\Omega^{\rho}_{\nu)} \, . \tag{2.35a}$$

---

[7] This name was introduced by Geracie et al. in 2015 [1], after previous authors had either given the form no specific name or, in one case, called it just the 'Coriolis form'. Since, as we will see later, in a certain sense it contains the Newtonian potential (and not just the effects of Coriolis forces), they decided to include the name 'Newton'.

Using the identity (2.34), this may be rewritten as

$$\Gamma^{\rho}_{\mu\nu} = v^{\rho}\partial_{(\mu}\tau_{\nu)} + \frac{1}{2}h^{\rho\sigma}(2\partial_{(\mu}h_{\nu)\sigma} - \partial_{\sigma}h_{\mu\nu}) + \frac{1}{2}T^{\rho}_{\mu\nu} - T_{(\mu\nu)}{}^{\rho} + \tau_{(\mu}\Omega^{\rho}_{\nu)} \ .$$

$$(2.35\text{b})$$

(ii) Conversely, given any two-form Ω and any tensor field $T \in \Gamma(TM \otimes \bigwedge^2 T^*M)$ satisfying the identity (2.34), Eq. (2.35) defines a Galilei connection on (M, τ, h), whose torsion and Newton–Coriolis form with respect to v are the given T and Ω, respectively.

Before proving the theorem, we are going to give a few remarks.

Remark 2.30 (i) Let us explicitly appreciate the 'classification' aspect of Theorem 2.29: it shows that and how a unit timelike vector field v induces a bijective correspondence between Galilei connections on the one hand and tensor fields Ω and T satisfying (2.34) (as well as their respective symmetry properties) on the other hand.

(ii) Consider the definition of a connection in terms of T and Ω according to part (ii) of the theorem. The identity (2.34) restricts only the torsion's timelike part on the first index $\tau_{\rho}T^{\rho}_{\mu\nu}$, leaving the spacelike part $P^{\rho}_{\lambda}T^{\lambda}_{\mu\nu}$ (equivalently $T_{\lambda\mu\nu}$) unconstrained.

From this point of view, a choice of unit timelike vector field v therefore induces an affine isomorphism between the affine space of Galilei connections on M and the vector space

$$\Gamma(\ker\tau \otimes \textstyle\bigwedge^2 T^*M) \oplus \Gamma(\textstyle\bigwedge^2 T^*M) \ , \qquad (2.36\text{a})$$

mapping a connection ∇ to the field with components

$$(P^{\rho}_{\lambda}T^{\lambda}_{\mu\nu} \ , \ \Omega_{\mu\nu}) \ . \qquad (2.36\text{b})$$

The first form (2.35a) of the connection coefficients from the theorem expresses the connection explicitly in terms of only these freely specifiable fields, i.e. it provides the inverse of this isomorphism explicitly.

Note that v is *not* an independent field in defining the connection. On the contrary, a choice of v only determines the isomorphism between the space of connections and the vector space (2.36a); the freedom in specifying the connection itself resides fully in this vector space.   △

The 'origin' in the space of connections singled out by a choice of v gets a specific name:

Definition 2.31 Let (M, τ, h) be a Galilei manifold and v a unit timelike vector field on it. The *special Galilei connection with respect to* v is the Galilei connection $\overset{v}{\nabla}$ having

(i) vanishing 'spatial torsion' $P^\rho_\lambda T^\lambda{}_{\mu\nu} = 0$ with respect to v, and
(ii) vanishing Newton–Coriolis form with respect to v.

This means that it has torsion $T^\rho{}_{\mu\nu} = v^\rho (\mathrm{d}\tau)_{\mu\nu}$, and its connection coefficients are

$$\overset{v}{\Gamma}{}^\rho_{\mu\nu} = v^\rho \partial_\mu \tau_\nu + \frac{1}{2} h^{\rho\sigma}(2\partial_{(\mu} h_{\nu)\sigma} - \partial_\sigma h_{\mu\nu}). \tag{2.37}$$

❀

In the case of absolute time, the classification theorem implies the following:

Corollary 2.32 *On a Galilei manifold with absolute time, the torsion-free Galilei connections are classified by their respective Newton–Coriolis forms (with respect to a choice of unit timelike vector field).* □

Note that this is different from the case of (pseudo-)Riemannian manifolds, where the metric-compatible connections are classified by their torsion and therefore there is a unique torsion-free connection, the Levi-Civita connection.

Now we are going to prove the classification theorem.

Proof of theorem 2.29.

(i) First, the identity (2.34) for the timelike torsion is of course just the statement of Proposition 2.20 (i).

Turning to the derivation of the classification formula (2.35), we decompose the connection coefficients on their upper index into their timelike and spacelike parts with respect to v, according to

$$\Gamma^\rho_{\mu\nu} = \delta^\rho_\lambda \Gamma^\lambda_{\mu\nu} = (v^\rho \tau_\lambda + P^\rho_\lambda)\Gamma^\lambda_{\mu\nu} = (v^\rho \tau_\lambda + h^{\rho\sigma} h_{\sigma\lambda})\Gamma^\lambda_{\mu\nu}. \tag{2.38}$$

The timelike part we obtain from the compatibility $\nabla \tau = 0$, which in components reads $0 = \partial_\mu \tau_\nu - \Gamma^\lambda_{\mu\nu}\tau_\lambda$, giving

$$\Gamma^\rho_{\mu\nu} = v^\rho \partial_\mu \tau_\nu + h^{\rho\sigma} h_{\sigma\lambda}\Gamma^\lambda_{\mu\nu}. \tag{2.39}$$

To compute the spacelike part, we will proceed in quite some parallel to the standard way of classifying metric connections on pseudo-Riemannian manifolds.

For any three vector fields X, Y, Z, by applying the Leibniz rule and the definition of the torsion we have

$$
\begin{aligned}
\underset{v}{h}(\nabla_X Y, Z) &= X\big(\underset{v}{h}(Y, Z)\big) - (\nabla_X \underset{v}{h})(Y, Z) - \underset{v}{h}(Y, \nabla_X Z) \\
&= X\big(\underset{v}{h}(Y, Z)\big) - (\nabla_X \underset{v}{h})(Y, Z) - \underset{v}{h}(Y, \nabla_Z X) \\
&\quad + \underset{v}{h}(Y, T(Z, X) + [Z, X]) \,.
\end{aligned}
\tag{2.40a}
$$

Cyclicly permuting the vector fields, from this we obtain

$$
\begin{aligned}
\underset{v}{h}(\nabla_Y Z, X) &= Y\big(\underset{v}{h}(Z, X)\big) - (\nabla_Y \underset{v}{h})(Z, X) - \underset{v}{h}(Z, \nabla_X Y) \\
&\quad + \underset{v}{h}(Z, T(X, Y) + [X, Y]) \,,
\end{aligned}
\tag{2.40b}
$$

$$
\begin{aligned}
\underset{v}{h}(\nabla_Z X, Y) &= Z\big(\underset{v}{h}(X, Y)\big) - (\nabla_Z \underset{v}{h})(X, Y) - \underset{v}{h}(X, \nabla_Y Z) \\
&\quad + \underset{v}{h}(X, T(Y, Z) + [Y, Z]) \,.
\end{aligned}
\tag{2.40c}
$$

Considering (2.40a) + (2.40b) − (2.40c) now yields

$$
\begin{aligned}
2\underset{v}{h}(\nabla_X Y, Z) &= X\big(\underset{v}{h}(Y, Z)\big) + Y\big(\underset{v}{h}(Z, X)\big) - Z\big(\underset{v}{h}(X, Y)\big) \\
&\quad - (\nabla_X \underset{v}{h})(Y, Z) - (\nabla_Y \underset{v}{h})(Z, X) + (\nabla_Z \underset{v}{h})(X, Y) \\
&\quad + \underset{v}{h}(Y, T(Z, X) + [Z, X]) + \underset{v}{h}(Z, T(X, Y) + [X, Y]) \\
&\quad - \underset{v}{h}(X, T(Y, Z) + [Y, Z]) \,.
\end{aligned}
\tag{2.41}
$$

To continue, we compute how to to express the covariant derivative $\nabla \underset{v}{h}$ of the covariant space metric in terms of ∇v, for which we introduce the shorthand notation $\Lambda_\mu{}^\nu := \nabla_\mu v^\nu$.

Lemma 2.33 *Consider a Galilei manifold (M, τ, h), and a unit timelike vector field v and a Galilei connection ∇ on it. Writing $\Lambda_\mu{}^\nu = \nabla_\mu v^\nu$, the covariant derivative of $\underset{v}{h}$ is*

$$
\nabla_\rho h_{\mu\nu} = -2\Lambda_{\rho(\mu}\tau_{\nu)}
\tag{2.42}
$$

(where indices have been lowered with $\underset{v}{h}$).

Proof We decompose $\nabla \underset{v}{h}$ on its second index into its spacelike and its timelike part with respect to v, namely

$$
\nabla_\rho h_{\mu\nu} = \tau_\mu v^\lambda \nabla_\rho h_{\lambda\nu} + h_{\mu\kappa} h^{\kappa\lambda} \nabla_\rho h_{\lambda\nu} \,.
\tag{2.43}
$$

The timelike part we obtain from

$$
0 = \nabla_\rho(v^\lambda h_{\lambda\nu}) = \Lambda_\rho{}^\lambda h_{\lambda\nu} + v^\lambda \nabla_\rho h_{\lambda\nu} = \Lambda_{\rho\nu} + v^\lambda \nabla_\rho h_{\lambda\nu} \,,
\tag{2.44a}
$$

and the spacelike one from

$$0 = \nabla_\rho \delta^\kappa_\nu = \nabla_\rho(P^\kappa_\nu + v^\kappa \tau_\nu) = \nabla_\rho(h^{\kappa\lambda}h_{\lambda\nu}) + \Lambda_\rho{}^\kappa \tau_\nu = h^{\kappa\lambda}\nabla_\rho h_{\lambda\nu} + \Lambda_\rho{}^\kappa \tau_\nu$$
$$(2.44b)$$

where we have used $\nabla\tau = 0$, $\nabla h = 0$. Inserting these into (2.43), we obtain (2.42). $\qquad\qquad\square$

Applying (2.41) to the coordinate vector fields $X = \partial_\mu$, $Y = \partial_\nu$, $Z = \partial_\sigma$ and using Lemma 2.33 to express $\nabla_v h$ in terms of $\Lambda = \nabla v$, we obtain

$$\begin{aligned}
h_{\sigma\lambda}\Gamma^\lambda_{\mu\nu} &= \frac{1}{2}(\partial_\mu h_{\nu\sigma} + \partial_\nu h_{\sigma\mu} - \partial_\sigma h_{\mu\nu}) + \frac{1}{2}(-\nabla_\mu h_{\nu\sigma} - \nabla_\nu h_{\sigma\mu} + \nabla_\sigma h_{\mu\nu}) \\
&\quad + \frac{1}{2}(T_{\nu\sigma\mu} + T_{\sigma\mu\nu} - T_{\mu\nu\sigma}) \\
&= \frac{1}{2}(2\partial_{(\mu}h_{\nu)\sigma} - \partial_\sigma h_{\mu\nu}) + \Lambda_{\mu(\nu}\tau_{\sigma)} + \Lambda_{\nu(\sigma}\tau_{\mu)} - \Lambda_{\sigma(\mu}\tau_{\nu)} \\
&\quad + \frac{1}{2}T_{\sigma\mu\nu} - T_{(\mu\nu)\sigma} \, .
\end{aligned}$$
$$(2.45)$$

Equation (2.45) gives us the desired expression for the spacelike part of the connection coefficients. Inserting it into (2.39), we obtain

$$\begin{aligned}
\Gamma^\rho_{\mu\nu} &= v^\rho \partial_\mu \tau_\nu + \frac{1}{2}h^{\rho\sigma}(2\partial_{(\mu}h_{\nu)\sigma} - \partial_\sigma h_{\mu\nu}) + \frac{1}{2}h^{\rho\sigma}T_{\sigma\mu\nu} - h^{\rho\sigma}T_{(\mu\nu)\sigma} \\
&\quad + h^{\rho\sigma}(\Lambda_{\mu(\nu}\tau_{\sigma)} + \Lambda_{\nu(\sigma}\tau_{\mu)} - \Lambda_{\sigma(\mu}\tau_{\nu)}) \\
&= v^\rho \partial_\mu \tau_\nu + \frac{1}{2}h^{\rho\sigma}(2\partial_{(\mu}h_{\nu)\sigma} - \partial_\sigma h_{\mu\nu}) + \frac{1}{2}P^\rho_\lambda T^\lambda{}_{\mu\nu} - T_{(\mu\nu)}{}^\rho \\
&\quad + h^{\rho\sigma}(\Lambda_{(\mu|\sigma|} - \Lambda_{\sigma(\mu)}\tau_{\nu)} \, ,
\end{aligned}$$
$$(2.46)$$

where we used that first lowering an index with $\underset{v}{h}$ and then raising it again with h amounts to spatially projecting it with P. Noting that the definition of the Newton–Coriolis form is $\Omega_{\mu\rho} = 2\Lambda_{[\mu\rho]} = \Lambda_{\mu\rho} - \Lambda_{\rho\mu}$, we may rewrite

$$(\Lambda_{(\mu|\sigma|} - \Lambda_{\sigma(\mu)}\tau_{\nu)} = \Omega_{(\mu|\sigma|}\tau_{\nu)} = \tau_{(\mu}\Omega_{\nu)\sigma} \, .$$
$$(2.47)$$

Thus, (2.46) yields the connection coefficients as stated in the first version of the classification formula (2.35a).

Finally, we note that due to the identity (2.34), the spatially projected torsion term in (2.35a) can be rewritten as

$$\begin{aligned}
\frac{1}{2}P^\rho_\lambda T^\lambda{}_{\mu\nu} &= \frac{1}{2}T^\rho{}_{\mu\nu} - \frac{1}{2}v^\rho \tau_\lambda T^\lambda{}_{\mu\nu} = \frac{1}{2}T^\rho{}_{\mu\nu} - \frac{1}{2}v^\rho(d\tau)_{\mu\nu} \\
&= \frac{1}{2}T^\rho{}_{\mu\nu} - v^\rho \partial_{[\mu}\tau_{\nu]} \, .
\end{aligned}$$
$$(2.48)$$

This shows that the two forms (2.35a) and (2.35b) of the connection coefficients are indeed equal.

(ii) Since T satisfies (2.34), the two forms (2.35a) and (2.35b) of the connection coefficients are equal. The proof that (2.35) defines a Galilei connection is covered in detail in Exercise 2.34. Its torsion being the given T is obvious from (2.35b) (consider $2\Gamma^{\rho}_{[\mu\nu]}$). That its Newton–Coriolis form is the given Ω requires some calculation, which we leave to Exercise 2.35. $\square$

Exercise 2.34 (*Existence of Galilei connections*)   Consider a Galilei manifold (M, τ, h) and a unit timelike vector field v on it. In this exercise, we are going to prove the first part of Theorem 2.29 (ii).

(a) Show that the 'special Galilei connection' $\overset{v}{\nabla}$, defined by having coefficients (2.37), is indeed a Galilei connection on (M, τ, h).

   Hint: Unfortunately, you have to ('just') do the calculation explicitly, using the connection coefficients. Showing $\overset{v}{\nabla}_{\mu}\tau_{\nu} = 0$ is easy; for $\overset{v}{\nabla}_{\mu}h^{\rho\sigma} = 0$ you need to apply the product rule 'backwards' a few times, and use the definitions of $h_{\mu\nu}$ and P^{μ}_{ν}.

(b) Show that (2.35a) defines a Galilei connection, where T is any tensor field in $\Gamma(TM \otimes \bigwedge^2 T^*M)$, and Ω is an arbitrary two-form. (We need not assume that T satisfy (2.34): this would only additionally ensure that the two forms (2.35a) and (2.35b) of the connection coefficients are equal, such that T really is the torsion.)

   Hint: Consider the difference to the special connection $\overset{v}{\nabla}$, and use Proposition 2.26. $\triangle$

Exercise 2.35 (*The Newton–Coriolis form of a general Galilei connection*)   Let (M, τ, h) be a Galilei manifold, v a unit timelike vector field on it, T any tensor field in $\Gamma(TM \otimes \bigwedge^2 T^*M)$, and Ω an arbitrary two-form.

   In this exercise, we are going to prove the remaining part of Theorem 2.29 (ii), namely that the Galilei connection defined by (2.35a) has Newton–Coriolis form Ω with respect to v.

(a) First show that the special Galilei connection $\overset{v}{\nabla}$ as defined by (2.37) has vanishing Newton–Coriolis form with respect to v, i.e. that $2h_{\rho[\nu}\overset{v}{\nabla}_{\mu]}v^{\rho} = 0$.

   Hint: Again, proceed by explicit calculation. You need to apply the product rule 'backwards' several times, and use the definitions of P^{μ}_{ν} and $h_{\mu\nu}$.

(b) Now writing the coefficients of the Galilei connection ∇ defined by (2.35a) as $\Gamma^{\rho}_{\mu\nu} = \overset{v}{\Gamma}^{\rho}_{\mu\nu} + S^{\rho}_{\mu\nu}$, the Newton–Coriolis form of ∇ has $\mu\nu$-components

$$2h_{\rho[\nu}\nabla_{\mu]}v^{\rho} = 2h_{\rho[\nu}\overset{v}{\nabla}^{(v)}_{\mu]}v^{\rho} + 2h_{\rho[\nu}S^{\rho}_{\mu]\sigma}v^{\sigma} = 2S_{[\nu\mu]\sigma}v^{\sigma}, \qquad (2.49)$$

where we used part (a). Insert the explicit form of S and show that you obtain $\Omega_{\mu\nu}$. $\triangle$

Now we are going to discuss how the objects associated to a choice of unit timelike vector field change when the vector field is changed.

Definition 2.36 On a Galilei manifold, any change from one unit timelike vector field to another has the form

$$v^\mu \mapsto \tilde{v}^\mu = v^\mu - w^\mu \tag{2.50}$$

for a spacelike vector field w. Such a change of unit timelike field is called a *Milne boost*.   ✾

Proposition 2.37 *Under a Milne boost* (2.50), *the spatial projector* P, *the covariant space metric* $\underset{v}{h}$ *and the Newton–Coriolis form (of some fixed Galilei connection) with respect to the vector field transform according to*

$$P \mapsto \tilde{P} = P + w \otimes \tau, \tag{2.51a}$$

$$\underset{v}{h} \mapsto \underset{\tilde{v}}{h} = \underset{v}{h} + w^\flat \otimes \tau + \tau \otimes w^\flat + |w|^2 \tau \otimes \tau, \tag{2.51b}$$

$$\Omega \mapsto \tilde{\Omega} = \Omega - \mathrm{d}w^\flat - \frac{1}{2}\mathrm{d}|w|^2 \wedge \tau + \underset{v}{h}(w, T(\cdot, \cdot)), \tag{2.51c}$$

where $w^\flat = \underset{v}{h}(w, \cdot)$ *is the one-form associated to* w *via* $\underset{v}{h}$, *and* $|w|^2 = {}^{(n)}h(w, w)$ *is the squared length of* w. *Written in components, the above formulae read*

$$P^\mu_\nu \mapsto P^\mu_\nu + w^\mu \tau_\nu \,, \tag{2.52a}$$

$$h_{\mu\nu} \mapsto h_{\mu\nu} + w_\mu \tau_\nu + \tau_\mu w_\nu + |w|^2 \tau_\mu \tau_\nu \,, \tag{2.52b}$$

$$\Omega_{\mu\nu} \mapsto \Omega_{\mu\nu} - 2\partial_{[\mu} w_{\nu]} - (\partial_{[\mu}|w|^2)\tau_{\nu]} + w_\rho T^\rho{}_{\mu\nu} \,, \tag{2.52c}$$

where the index on w_μ *has been lowered with* $\underset{v}{h}$.

Proof The transformation behaviour of the projector follows directly from its definition. For the covariant space metric and the Newton–Coriolis form, some calculation has to be done, which we leave to Exercise 2.38.   □

Exercise 2.38 (*Milne boosts*)
(a) Show that under the Milne boost (2.50), the spatial projector P along v changes according to (2.52a), i.e. that the right-hand side of this formula are the components of the projector with respect to $\tilde{v}$.
(b) Show that under the Milne boost (2.50), the covariant space metric $\underset{v}{h}$ with respect to v changes according to (2.52b), i.e. that the right-hand side of this formula are the components of the covariant space metric $\underset{\tilde{v}}{h}$ with respect to $\tilde{v}$.
   Hint: Show by direct computation that the right-hand side satisfies the two defining equations for $\underset{v}{h}$. *Be careful: you have to use the projector with respect to* $\tilde{v}$!

(c) Let ∇ be a Galilei connection. Show that under the Milne boost (2.50), the Newton–Coriolis form Ω of ∇ with respect to v changes according to (2.52c), i.e. show that if Ω is the Newton–Coriolis form of ∇ with respect to v, then the right-hand side of (2.52c) are the components of the Newton–Coriolis form of ∇ with respect to $\tilde{v}$.

Hint: You have to be careful: in the definition of the Newton–Coriolis form, both the vector field v and the covariant space metric used to lower the index before antisymmetrisation change under the Milne boost. By inserting these and computing step by step, you should arrive at the desired result. You will need to apply the product rule 'backwards' once, and to use Lemma 2.33 to express $\nabla\!{}_v h$ in terms of ∇v. You will also encounter a term of the form $2w_\lambda \nabla_\mu w^\lambda$, which you can without computation rewrite as $\nabla_\mu |w|^2 = \partial_\mu |w|^2$, using that the connection induced on the spacelike distribution is metric with respect to the induced metric ${}^{(n)}h$ (Construction 2.22).   △

At this point, the transformation behaviour of $\underset{v}{h}$ and Ω under Milne boosts seems to appear 'out of the blue'. Later, we will see how to actually *derive* it.

2.4 Newtonian Connections

Consider a Galilei manifold with absolute time, and a torsion-free Galilei connection on it. We want to count the number of algebraically independent components of the connection's curvature tensor (at any point), and compare this to the (pseudo-)Riemannian case.

Recall that for the Levi-Civita connection of an n-dimensional (pseudo-)Riemannian manifold, the (Riemannian) curvature tensor has $\frac{n^2(n^2-1)}{12}$ independent components. Since the curvature tensor of a Galilei connection has less symmetries than in the Riemannian case, it has more independent components:

Proposition 2.39 *Let (M, τ, h) be an $(n+1)$-dimensional Galilei manifold with absolute time, and ∇ a torsion-free Galilei connection on it. At any point of M, the number of algebraically independent components of the curvature tensor of ∇ (i.e. the dimension of the vector space of tensors with the algebraic symmetries of the curvature tensor) is*

$$\frac{n^2(n+1)(n+5)}{12} = \frac{(n+1)^2((n+1)^2-1)}{12} + \frac{n(n^2-1)}{6}, \tag{2.53}$$

i.e. $\frac{n(n^2-1)}{6}$ more than that for an $(n+1)$-dimensional (pseudo-)Riemannian manifold.

Proof The first index of the curvature tensor is spacelike (Proposition 2.20 (ii)),

$$\tau_\mu R^\mu{}_{\nu\rho\sigma} = 0. \tag{2.54a}$$

Furthermore, the curvature tensor has the following symmetries: it is antisymmetric in its last two indices,

$$R^{\mu}{}_{\nu\rho\sigma} = -R^{\mu}{}_{\nu\sigma\rho}\,, \tag{2.54b}$$

antisymmetric in its first two indices when the second is raised (Proposition 2.20 (ii)),

$$R^{\mu\nu}{}_{\rho\sigma} = -R^{\nu\mu}{}_{\rho\sigma}\,, \tag{2.54c}$$

and due to vanishing torsion satisfies the Bianchi identity

$$R^{\mu}{}_{[\nu\rho\sigma]} = 0. \tag{2.54d}$$

Due to antisymmetry in the last two indices, the Bianchi identity can also be stated as the vanishing of the cyclic sum over the last three indices,

$$R^{\mu}{}_{\nu\rho\sigma} + R^{\mu}{}_{\rho\sigma\nu} + R^{\mu}{}_{\sigma\nu\rho} = 0. \tag{2.54e}$$

We now focus on a single point $p \in M$ and want to analyse these symmetries in a Galilei basis $(e_A) = (e_t, e_a)$ of T_pM. The first three symmetries (2.54) of the curvature then take the form

$$R^{t}{}_{\nu\rho\sigma} = 0, \tag{2.55a}$$
$$R_{a\nu\rho\sigma} = -R_{a\nu\sigma\rho}\,, \tag{2.55b}$$
$$R_{ab\rho\sigma} = -R_{ba\rho\sigma}\,, \tag{2.55c}$$

where we have lowered the first index with δ_{ab} when it is spacelike.

The purely spacelike part R_{abcd} of the curvature tensor now has precisely the symmetries of the curvature tensor of an n-dimensional Riemannian manifold (in fact, according to Construction 2.24 it *is* the Riemannian curvature of the spatial leaves), so it has $\frac{n^2(n^2-1)}{12}$ independent components. (Note that the 'symmetry in pairs' $R_{abcd} = R_{cdab}$ follows from the Bianchi identity and the antisymmetries.)

Thus, we still need to count the number of independent mixed timelike-spacelike components of the curvature, which have the form $R_{a\nu\rho\sigma} = -R_{a\nu\sigma\rho}$ with at least one of the indices ν, ρ, σ being t. On the one hand, this includes components of the form $R_{at\rho\sigma}$. For those, we have n independent choices for the value of a, and $\frac{(n+1)n}{2}$ choices for the antisymmetric pair $\rho\sigma$; thus these are $n \cdot \frac{(n+1)n}{2}$ independent components. On the other hand, we have the components R_{abtc}. Here for the antisymmetric pair ab, we have $\frac{n(n-1)}{2}$ choices, and n choices for c, meaning those give us $\frac{n(n-1)}{2} \cdot n$ independent components.

In the above counting, we have however not yet taken the Bianchi identity into account, which tells us that $R_{at\rho\sigma} + $ (cyclic terms in $t, \rho, \sigma) = 0$. This is a non-trivial requirement (i.e. one which does not follow from the already considered symmetries) for either all four indices being different, or for one of ρ, σ being a. The former case gives $n \cdot \binom{n-1}{2}$ equations (n choices for a, then two values ρ, σ from the remaining $n - 1$); the latter amounts to $0 = R_{atab} + R_{aabt} + R_{abta} = R_{atab} + R_{abta}$, which are $n(n - 1)$ independent equations.

Combining everything, the number of independent components of the curvature tensor at p is

$$\underbrace{\frac{n^2(n^2 - 1)}{12}}_{R_{abcd}} + \underbrace{n \cdot \frac{(n + 1)n}{2}}_{R_{at\rho\sigma}} + \underbrace{\frac{n(n - 1)}{2} \cdot n}_{R_{abtc}} - \underbrace{n \cdot \binom{n - 1}{2} - n(n - 1)}_{\text{Bianchi}}, \quad (2.56)$$

which by a little calculation can be shown to be equal to the number given in (2.53). $\square$

We now want to introduce a requirement on the curvature that reduces the number of independent components to as many as in the (pseudo-)Riemannian case. The only symmetry holding in the (pseudo-)Riemannian case but absent above was the 'symmetry in pairs' for the mixed timelike-spacelike curvature components.

Definition 2.40 A *Newtonian connection* on a Galilei manifold (M, τ, h) with absolute time is a torsion-free Galilei connection whose curvature tensor satisfies the additional symmetry

$$R^{\mu}{}_{\rho}{}^{\nu}{}_{\sigma} = R^{\nu}{}_{\sigma}{}^{\mu}{}_{\rho} . \quad (2.57)$$

A *Newtonian manifold* is a Galilei manifold with absolute time with a Newtonian connection on it. ❀

Theorem 2.41 *The number of algebraically independent components of the curvature tensor of an $(n + 1)$-dimensional Newtonian manifold is*

$$\frac{(n + 1)^2((n + 1)^2 - 1)}{12}, \quad (2.58)$$

as for an $(n + 1)$-dimensional (pseudo-)Riemannian manifold.

Proof In a Galilei basis, the additional symmetries introduced by the condition of being Newtonian read

$$R_{atbt} = R_{btat} \text{ for } a \neq b, \quad (2.59a)$$

$$R_{atbc} = R_{bcat} \text{ for } a, b, c \text{ all different} \quad (2.59b)$$

(for $a = b$, the second one follows from the Bianchi identity). The first case gives us $\binom{n}{2}$ equations, and the second one $\binom{n}{3}$. Together, these are

$$\binom{n}{2} + \binom{n}{3} = \frac{n(n - 1)}{2} + \frac{n(n - 1)(n - 2)}{6} = \frac{n(n - 1)(n + 1)}{6} = \frac{n(n^2 - 1)}{6} \quad (2.60)$$

additional requirements for the curvature, which reduces the number of components from (2.53) to the (pseudo-)Riemannian value. $\square$

The condition of a Galilei connection being Newtonian can be related to its Newton–Coriolis form:

Lemma 2.42 *Let (M, τ, h) be a Galilei manifold with absolute time, ∇ be a torsion-free Galilei connection on it, and Ω be its Newton–Coriolis form with respect to any unit timelike vector field. Then the curvature tensor of ∇ satisfies*

$$R^{\mu}{}_{\rho}{}^{\nu}{}_{\sigma} - R^{\nu}{}_{\sigma}{}^{\mu}{}_{\rho} = (\mathrm{d}\Omega)^{\mu\nu}{}_{(\rho}\tau_{\sigma)} \; . \tag{2.61}$$

Proof We denote the unit timelike vector field by v, and its covariant derivative by $\nabla v =: \Lambda$. Then the definition of Ω reads $\Omega_{\mu\nu} = 2\Lambda_{[\mu\nu]}$. Due to torsion-freeness, i.e. $\Gamma^{\lambda}_{[\mu\nu]} = \frac{1}{2}T^{\lambda}{}_{\mu\nu} = 0$, we can thus express the exterior derivative as

$$\begin{aligned}
(\mathrm{d}\Omega)_{\mu\nu\rho} &= 3\,\partial_{[\mu}\Omega_{\nu\rho]} \\
&= 3(\nabla_{[\mu}\Omega_{\nu\rho]} + \Gamma^{\lambda}_{[\mu\nu}\Omega_{|\lambda|\rho]} + \Gamma^{\lambda}_{[\mu\rho}\Omega_{\nu]\lambda}) \\
&= 6\,\nabla_{[\mu}\Lambda_{\nu\rho]} \; .
\end{aligned} \tag{2.62}$$

Now in general, for any torsion-free connection and any one-form α, we have the identity

$$2\nabla_{[\mu}\nabla_{\nu]}\alpha_{\rho} = -R^{\lambda}{}_{\rho\mu\nu}\alpha_{\lambda} \tag{2.63}$$

(with torsion, we would have an additional $-T^{\lambda}{}_{\mu\nu}\nabla_{\lambda}\alpha_{\rho}$ on the right-hand side). Applying this to a tensor product $\alpha_{\rho}\beta_{\sigma}$, we obtain

$$\begin{aligned}
2\nabla_{[\mu}\nabla_{\nu]}(\alpha_{\rho}\beta_{\sigma}) &= 2(\nabla_{[\mu}\nabla_{\nu]}\alpha_{\rho})\beta_{\sigma} + 2\alpha_{\rho}\nabla_{[\mu}\nabla_{\nu]}\beta_{\sigma} \\
&= -R^{\lambda}{}_{\rho\mu\nu}\alpha_{\lambda}\beta_{\sigma} - R^{\lambda}{}_{\sigma\mu\nu}\alpha_{\rho}\beta_{\lambda} \; .
\end{aligned} \tag{2.64}$$

Since the covariant space metric $h_{\rho\sigma}$ may be decomposed as a sum of tensor products of one-forms, we obtain

$$\nabla_{[\mu}\nabla_{\nu]}h_{\rho\sigma} = -h_{\lambda(\rho}R^{\lambda}{}_{\sigma)\mu\nu} \; , \tag{2.65}$$

which with Lemma 2.33 takes the form

$$R_{(\rho\sigma)\mu\nu} = 2\nabla_{[\mu}\Lambda_{\nu](\rho}\tau_{\sigma)} \; . \tag{2.66}$$

Contracting this with $6v^{\sigma}$ and antisymmetrising in ρ, μ, ν, we obtain

$$3R_{[\rho|\sigma|\mu\nu]}v^{\sigma} = 6\nabla_{[\mu}\Lambda_{\nu\rho]} + 6\tau_{[\rho}\nabla_{\mu}\Lambda_{\nu]\sigma}v^{\sigma} . \tag{2.67}$$

Now due to $\Lambda_{\nu\sigma}v^\sigma = \Lambda_\nu{}^\kappa h_{\kappa\sigma}v^\sigma = 0$, we have

$$
\begin{aligned}
\nabla_{[\mu}\Lambda_{\nu]\sigma}v^\sigma &= \nabla_{[\mu}(\Lambda_{\nu]\sigma}v^\sigma) - \Lambda_{[\nu|\sigma|}\nabla_{\mu]}v^\sigma \\
&= -\Lambda_{[\nu|\sigma|}\Lambda_{\mu]}{}^\sigma \\
&= -\frac{1}{2}(\Lambda_{\nu\sigma}\Lambda_\mu{}^\sigma - \Lambda_{\mu\sigma}\Lambda_\nu{}^\sigma) = 0.
\end{aligned}
\tag{2.68}
$$

Using this and (2.67) to rewrite (2.62), we arrive at

$$
\begin{aligned}
(\mathrm{d}\Omega)_{\mu\nu\rho} &= 3R_{[\rho|\lambda|\mu\nu]}v^\lambda \\
&= (R_{\rho\lambda\mu\nu} + R_{\mu\lambda\nu\rho} + R_{\nu\lambda\rho\mu})v^\lambda,
\end{aligned}
\tag{2.69}
$$

where we have used antisymmetry of the curvature in the last two indices to rewrite the antisymmetrisation as a cyclic sum. Now using the Bianchi identity, we can rewrite the first and second terms on the right-hand side of (2.69) as

$$
R_{\rho\lambda\mu\nu} = -R_{\rho\mu\nu\lambda} - R_{\rho\nu\lambda\mu} ,
\tag{2.70a}
$$
$$
R_{\mu\lambda\nu\rho} = -R_{\mu\nu\rho\lambda} - R_{\mu\rho\lambda\nu} .
\tag{2.70b}
$$

By decomposing the second index of R into its space- and timelike parts, we obtain

$$
\begin{aligned}
R_{(\mu\nu)\rho\sigma} &= (P_{(\nu}^\kappa + v^\kappa\tau_{(\nu})R_{\mu)\kappa\rho\sigma} \\
&= h_{\lambda(\mu}h_{|\kappa|\nu)}R^{\lambda\kappa}{}_{\rho\sigma} + v^\kappa\tau_{(\nu}R_{\mu)\kappa\rho\sigma} \\
&= h_{\lambda\mu}h_{\kappa\nu}R^{(\lambda\kappa)}{}_{\rho\sigma} + v^\kappa\tau_{(\nu}R_{\mu)\kappa\rho\sigma} \\
&= v^\kappa\tau_{(\mu}R_{\nu)\kappa\rho\sigma} ,
\end{aligned}
\tag{2.71}
$$

where in the last step we used antisymmetry of the spacelike part of the first two indices of the curvature tensor (Proposition 2.20 (ii)). Using this, we can rewrite both terms on the right-hand side of (2.70a) and the first term on the right-hand side of (2.70b) as

$$
-R_{\rho\mu\nu\lambda} = R_{\mu\rho\nu\lambda} - 2v^\kappa\tau_{(\rho}R_{\mu)\kappa\nu\lambda} ,
\tag{2.72a}
$$
$$
-R_{\rho\nu\lambda\mu} = R_{\nu\rho\lambda\mu} - 2v^\kappa\tau_{(\rho}R_{\nu)\kappa\lambda\mu} ,
\tag{2.72b}
$$
$$
-R_{\mu\nu\rho\lambda} = R_{\nu\mu\rho\lambda} - 2v^\kappa\tau_{(\mu}R_{\nu)\kappa\rho\lambda} .
\tag{2.72c}
$$

Using (2.70) and (2.72) in (2.69), we obtain

$$
\begin{aligned}
(\mathrm{d}\Omega)_{\mu\nu\rho} &= (R_{\mu\rho\nu\lambda} + R_{\nu\rho\lambda\mu} + R_{\nu\mu\rho\lambda} - R_{\mu\rho\lambda\nu} + R_{\nu\lambda\rho\mu} \\
&\quad - 2v^\kappa\tau_{(\rho}R_{\mu)\kappa\nu\lambda} - 2v^\kappa\tau_{(\rho}R_{\nu)\kappa\lambda\mu} - 2v^\kappa\tau_{(\mu}R_{\nu)\kappa\rho\lambda})v^\lambda \\
&= (2R_{\mu\rho\nu\lambda} - 2R_{\nu\lambda\mu\rho} - 2v^\kappa\tau_{(\rho}R_{\mu)\kappa\nu\lambda} \\
&\quad + 2v^\kappa\tau_{(\rho}R_{\nu)\kappa\mu\lambda} - 2v^\kappa\tau_{(\mu}R_{\nu)\kappa\rho\lambda})v^\lambda,
\end{aligned}
\tag{2.73}
$$

where we have used the Bianchi identity and antisymmetry in the last two indices to combine some terms. Raising μ and ν in this equation, we get

$$
\begin{aligned}
(d\Omega)^{\mu\nu}{}_\rho &= (2R^\mu{}_\rho{}^\nu{}_\lambda - 2R^\nu{}_\lambda{}^\mu{}_\rho - v^\kappa\tau_\rho R^\mu{}_\kappa{}^\nu{}_\lambda + v^\kappa\tau_\rho R^\nu{}_\kappa{}^\mu{}_\lambda)v^\lambda \\
&= \underbrace{(R^\mu{}_\kappa{}^\nu{}_\lambda - R^\nu{}_\lambda{}^\mu{}_\kappa)}_{=:M^\mu{}_\kappa{}^\nu{}_\lambda}(2\delta^\kappa_\rho - v^\kappa\tau_\rho)v^\lambda \\
&= M^\mu{}_\kappa{}^\nu{}_\lambda(\delta^\kappa_\rho + P^\kappa_\rho)v^\lambda.
\end{aligned}
\tag{2.74}
$$

Due to antisymmetry of $d\Omega$, from this we also obtain

$$
\begin{aligned}
(d\Omega)^{\mu\nu}{}_\sigma &= -(d\Omega)^{\nu\mu}{}_\sigma \overset{2.74}{=} -M^\nu{}_\kappa{}^\mu{}_\lambda(\delta^\kappa_\sigma + P^\kappa_\sigma)v^\lambda \\
&= M^\mu{}_\lambda{}^\nu{}_\kappa(\delta^\kappa_\sigma + P^\kappa_\sigma)v^\lambda = M^\mu{}_\kappa{}^\nu{}_\lambda(\delta^\lambda_\sigma + P^\lambda_\sigma)v^\kappa,
\end{aligned}
\tag{2.75}
$$

where we have used the definition of M in the second to last step and renamed contracted indices in the last step. Now multiplying (2.74) with τ_σ and (2.75) with τ_ρ, we get the two equations

$$
(d\Omega)^{\mu\nu}{}_\rho\tau_\sigma = M^\mu{}_\kappa{}^\nu{}_\lambda(\delta^\kappa_\rho + P^\kappa_\rho)(\delta^\lambda_\sigma - P^\lambda_\sigma),
\tag{2.76a}
$$

$$
(d\Omega)^{\mu\nu}{}_\sigma\tau_\rho = M^\mu{}_\kappa{}^\nu{}_\lambda(\delta^\kappa_\rho - P^\kappa_\rho)(\delta^\lambda_\sigma + P^\lambda_\sigma).
\tag{2.76b}
$$

Adding these and dividing by 2, we arrive at

$$
\begin{aligned}
(d\Omega)^{\mu\nu}{}_{(\rho}\tau_{\sigma)} &= M^\mu{}_\kappa{}^\nu{}_\lambda(\delta^\kappa_\rho\delta^\lambda_\sigma - P^\kappa_\rho P^\lambda_\sigma) \\
&= M^\mu{}_\rho{}^\nu{}_\sigma - M^{\mu\kappa\nu\lambda}h_{\rho\kappa}h_{\sigma\lambda}.
\end{aligned}
\tag{2.77}
$$

Due to symmetry in pairs of the purely spacelike part of the curvature tensor, we have $M^{\mu\kappa\nu\lambda} = 0$, and the proof is finished. $\square$

Theorem 2.43 *A torsion-free Galilei connection is Newtonian if and only if its Newton–Coriolis form with respect to any unit timelike vector field (and then all such fields) is closed.*

Proof If $d\Omega = 0$, then by Lemma 2.42 the connection is Newtonian. Conversely, if it is Newtonian, we have $(d\Omega)^{\mu\nu}{}_{(\sigma}\tau_{\rho)} = 0$. This implies $(d\Omega)^{\mu\nu\sigma} = 0$, i.e. vanishing of the purely spacelike part of $d\Omega$, and $(d\Omega)^{\mu\nu}{}_\sigma v^\sigma = 0$ for the unit timelike field v. Due to the antisymmetry of $d\Omega$, this implies $d\Omega = 0$. $\square$

Locally, we can use this to characterise Newtonian connections in still another way:

Theorem 2.44 *Let (M, τ, h) be a Galilei manifold with absolute time, and ∇ a torsion-free Galilei connection on it. Then ∇ is Newtonian if and only if for any point $p \in M$, there exists a unit timelike vector field $\tilde{v}$ defined on an open neighbourhood of p such that the Newton–Coriolis form of ∇ with respect to $\tilde{v}$ vanishes.*

Put differently: the Newtonian connections are precisely those torsion-free Galilei connections which are locally special.

Proof If we have a locally defined unit timelike vector field $\tilde{v}$ such that the Newton–Coriolis form of ∇ with respect to $\tilde{v}$ *vanishes*, then in particular it is closed; hence, by Theorem 2.43, ∇ is Newtonian on the domain of $\tilde{v}$.

Conversely, let ∇ be a Newtonian connection, and let v be *any* unit timelike vector field. By Theorem 2.43, the Newton–Coriolis form Ω of ∇ with respect to v is closed, $\mathrm{d}\Omega = 0$. By the Poincaré lemma, this means that it is locally exact, i.e. locally there is a one-form a such that $\Omega = -\mathrm{d}a$. We need to find a spacelike vector field w such that the Newton–Coriolis form $\tilde{\Omega}$ with respect to the Milne-boosted field $\tilde{v} = v - w$ vanishes. By (2.51c), $\tilde{\Omega}$ is given by

$$\tilde{\Omega} = \Omega - \mathrm{d}w^{\flat} - \frac{1}{2}\mathrm{d}|w|^2 \wedge \tau, \tag{2.78}$$

where $w^{\flat} = \underset{v}{h}(w, \cdot)$ and $|w|^2 = {}^{(n)}h(w, w)$. Hence, we need w to solve the equation

$$0 = -\mathrm{d}a - \mathrm{d}w^{\flat} - \frac{1}{2}\mathrm{d}|w|^2 \wedge \tau. \tag{2.79}$$

Due to (M, τ, h) having absolute time, we have $\mathrm{d}|w|^2 \wedge \tau = \mathrm{d}(|w|^2\tau)$; so a sufficient condition for this (which by the Poincaré lemma is also locally necessary) is

$$a + w^{\flat} + \frac{1}{2}|w|^2\tau = -\mathrm{d}\chi \tag{2.80a}$$

for some function χ. In components, this reads

$$a_\mu + h_{\mu\nu}w^\nu + \frac{1}{2}|w|^2\tau_\mu = -\partial_\mu\chi\,, \tag{2.80b}$$

where we have made the 'index lowering' on $w^{\flat}$ explicit. This equation is equivalent to its timelike and spacelike parts with respect to v being satisfied, i.e. to

$$a(v) + \frac{1}{2}|w|^2 = -\mathrm{d}\chi(v) \tag{2.81a}$$

and

$$h^{\mu\nu}a_\nu + w^\mu = -h^{\mu\nu}\partial_\nu\chi. \tag{2.81b}$$

The spacelike equation determines w as $w = -h(\mathrm{d}\chi + a, \cdot)$. Using this, the timelike equation becomes

$$a(v) + \frac{1}{2}h(\mathrm{d}\chi + a, \mathrm{d}\chi + a) = -\mathrm{d}\chi(v). \tag{2.82}$$

Note that if we can argue for this equation to have a local solution χ, we are finished.

Now, due to absolute time $\mathrm{d}\tau = 0$, locally we have $\tau = \mathrm{d}t$ for a function t, which we extend to local coordinates (t, x^a). Due to $\tau_\mu h^{\mu\nu} = 0$, in these coordinates we have $h = h^{ab}\partial_a \otimes \partial_b$, and due to $\tau(v) = 1$, we have $v = \partial_t + v^a\partial_a$. Thus, in these coordinates (2.82) becomes

$$a_t + a_a v^a + \frac{1}{2}h^{ab}(a_a + \partial_a\chi)(a_b + \partial_b\chi) = -\partial_t\chi - v^a\partial_a\chi. \tag{2.83}$$

This is a Hamilton–Jacobi equation

$$-\frac{\partial\chi(t, \vec{x})}{\partial t} = H\left(t, \vec{x}, \frac{\partial\chi(t, \vec{x})}{\partial\vec{x}}\right) \tag{2.84a}$$

with Hamiltonian

$$H(t, \vec{x}, \vec{p}) = a_t + a_a v^a + \frac{1}{2}h^{ab}(a_a + p_a)(a_b + p_b) + v^a p_a, \tag{2.84b}$$

where on the right-hand side the field components h^{ab}, v^a, a_μ are to be evaluated at $(t, \vec{x})$. Since τ, h, a and v are sufficiently regular, the corresponding Hamiltonian equations of motion admit (local) solutions, which is equivalent to the Hamilton–Jacobi equation admitting a local solution. $\square$

References

1. Geracie, M., Prabhu, K., Roberts, M.M.: Curved non-relativistic spacetimes, Newtonian gravitation and massive matter. J. Math. Phys. **56**(10), 103505 (2015). https://doi.org/10.1063/1.4932967
2. Schwartz, P.K.: The classification of general affine connections in Newton–Cartan geometry: towards metric-affine Newton–Cartan gravity. Class. Quantum Gravity **42**(1), 015010 (2025). https://doi.org/10.1088/1361-6382/ad922f

Classical Newton–Cartan Gravity 3

Abstract

In this chapter, we will use the language of Galilei manifolds and connections developed previously to introduce and discuss 'classical' Newton–Cartan gravity, i.e. the version of the theory presented by Künzle [6], Ehlers [2,3], and Malament [7, Chap. 4]. We will see how to recover the usual formulation of Newtonian gravity, and explore the connection to GR.

3.1 Formulation of the Theory

The essential content of Newton–Cartan gravity may be formulated as follows.

Axioms 3.1 (*Axioms for Newton–Cartan gravity*)
 (i) Spacetime is a Newtonian manifold (M, τ, h, ∇),
 (ii) ideal clocks measure time as defined by τ, and ideal rods measure spatial lengths as defined by the metric $^{(n)}h$ induced by h on spacelike vectors,
(iii) free test particles move on timelike geodesics of ∇, and
(iv) the *Newton–Cartan field equation*

$$R_{\mu\nu} = 4\pi G \rho \tau_\mu \tau_\nu \tag{3.1}$$

holds, where $R_{\mu\nu}$ are the components of the Ricci tensor of ∇, G is the gravitational constant, and ρ is the mass density. ✿

Note that as in GR, in Newton–Cartan gravity 'gravitational' effects are encoded into spacetime geometry, namely the connection. Differently to standard Newtonian gravity, and exactly as in GR, gravity is not a force, which would mean it leads to deviations from inertial motion; instead the distribution of matter is related to the

P. K. Schwartz, *Newton–Cartan Gravity*, Lecture Notes in Physics 1044,
https://doi.org/10.1007/978-3-032-03967-5_3

curvature of the connection, thus influencing the very notion of free/inertial motion itself, leading to those effects we call 'gravity'.[1]

Theorem 3.2 *Any* $(3 + 1)$*-dimensional Galilei manifold with absolute time with a torsion-free Galilei connection satisfying the Newton–Cartan field equation is spatially flat.*

Proof The field equation implies that the Ricci tensor vanishes on spacelike vectors. According to Construction 2.24, this means that all spatial leaves are Ricci-flat. Since they are 3-dimensional, this implies that they are flat as Riemannian manifolds. $\square$

For analysing timelike geodesics of ∇ (i.e. the worldlines of freely falling test particles), the following result is helpful.

Proposition 3.3 *Let* (M, τ, h) *be a Galilei manifold with absolute time, and* ∇ *a Galilei connection on it. Absolute time t as defined by* τ *(i.e. any local function t such that* $\tau = \mathrm{d}t$*) is an affine parameter for all timelike geodesics of* ∇*.*

Proof Let $\gamma(\lambda)$ be any affinely parametrised geodesic of ∇, i.e. $\nabla_{\gamma'(\lambda)}\gamma'(\lambda) = 0$. This implies $\nabla_{\gamma'(\lambda)}(\tau(\gamma'(\lambda))) = (\nabla_{\gamma'(\lambda)}\tau)(\gamma'(\lambda)) + \tau(\nabla_{\gamma'(\lambda)}\gamma'(\lambda)) = 0$, i.e. that $\tau(\gamma') =: A$ is constant along γ. Therefore, we have

$$t(\gamma(\lambda_{\mathrm{f}})) - t(\gamma(\lambda_{\mathrm{i}})) = \int_{\lambda_{\mathrm{i}}}^{\lambda_{\mathrm{f}}} \mathrm{d}\lambda \; \tau(\gamma'(\lambda)) = (\lambda_{\mathrm{f}} - \lambda_{\mathrm{i}})A, \tag{3.2}$$

i.e. absolute time t is affinely related to λ and therefore an affine parameter itself. $\square$

Notation 3.4 Unless otherwise specified, we will from now on parametrise all timelike curves by absolute time, i.e. such that $\tau(\dot{\gamma}) = 1$.   ✿

Using freely moving test particles, whose worldlines we know to be geodesics in Newton–Cartan gravity, we may operationally define the notion of *non-rotating directions* from the point of view of an observer on some specified worldline:

Exercise 3.5 (*Operational definition of non-rotating directions*)   For an observer moving on an arbitrary timelike worldline there is, a priori, no notion of 'non-rotating' directions. This is the same in GR as in Newton–Cartan gravity. In this exercise, we will explore a physically sensible method to define operationally what is meant by 'the same direction' over the course of time, and see how this notion is realised mathematically.

---

[1] For the sake of simplicity, we have only axiomatised the motion of test particles—a concept which is, of course, an idealisation without an actual counterpart in reality. One can also couple more realistic matter models to Newton–Cartan gravity, such as, e.g., (Newtonian) ideal fluids.

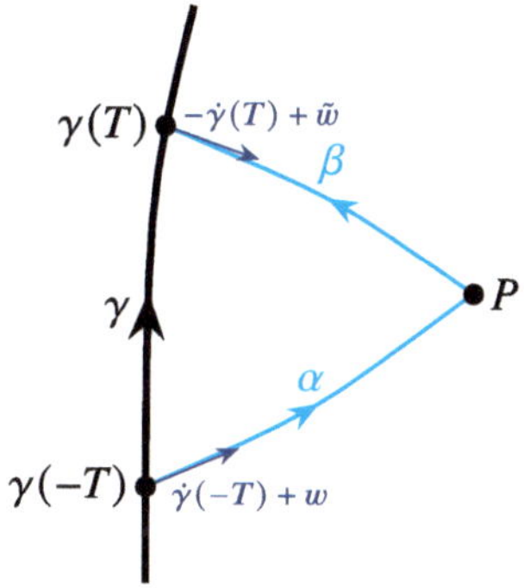

Fig. 3.1 The definition of non-rotating directions via a 'bouncing ball'

The method works as follows.[2] Imagine an observer moving through spacetime. At some point in time, the observer throws away a 'ball' with initial speed $|w|$ in some initial direction. The ball moves freely, and eventually gets 'reflected' back towards the observer in such a way that when it arrives again at the observer, its arrival speed is again $|w|$. The direction from which the ball arrives is then to be called 'the same direction' as the initial direction, at this later time. We are going to show that to linear order in the 'travelling time' of the ball, this 'same direction' is uniquely determined.

More precisely, the situation is as follows (see Fig. 3.1). We work in a Galilei manifold (M, τ, h) with absolute time, with a torsion-free Galilei connection ∇. We call the observer's worldline γ, which may be any timelike curve in M. At time $-T$, the ball is thrown with initial velocity $w \in \ker \tau|_{\gamma(-T)}$ (as seen from γ), and then moves along a freely falling worldline (i.e. geodesic of ∇) α. We want the ball to arrive again on γ at time T with velocity $-\tilde{w} \in \ker \tau|_{\gamma(T)}$, where $|w| = |\tilde{w}|$, moving on a geodesic β that crosses α (in some event P, where the 'reflection' happened). We will show below that this procedure determines $\tilde{w}$ to linear order in T, and that the map $w \mapsto \tilde{w}$ is given by parallel transport[3] along γ.

For simplicity, we parametrise all timelike curves by absolute time t, and work in coordinates (t, x^a) where the x^a are coordinates on the spatial leaves such that the observer's worldline γ corresponds to $x^a = 0$ (i.e. we have $\gamma^t(t) = t$, $\gamma^a(t) = 0$). We also use the notation $N := \gamma(0)$.

(a) The first worldline $\alpha(t)$ of the ball has to satisfy the initial value problem

$$\nabla_{\dot{\alpha}}\dot{\alpha} = 0, \quad \alpha(-T) = \gamma(-T), \quad \dot{\alpha}(-T) = \dot{\gamma}(-T) + w. \tag{3.3}$$

---

[2] This is an adaptation of Pirani's and Synge's 'bouncing photon' method for operationally defining non-rotation in GR [8], [11, Chap. III §8].

[3] In GR, the transport of non-rotating directions is not given by parallel transport with respect to the Levi-Civita connection, but instead by so-called *Fermi–Walker transport*. This is parallel transport with respect to the Fermi–Walker connection, which is defined by application of the Levi-Civita connection followed by orthogonal projection onto the orthogonal complement of $\dot{\gamma}$.

By expressing this in the coordinates (t, x^a) and Taylor-expanding $\alpha(t)$ around $t = -T$ to second order, determine $\alpha(t)$ approximately, for $t = \mathrm{O}(T)$.

Hint: You only need to consider the spacelike component of the geodesic equation, which reads $\ddot{\alpha}^a(t) + \Gamma^a_{\mu\nu}(\alpha(t))\dot{\alpha}^\mu(t)\dot{\alpha}^\nu(t) = 0$. From this, you can obtain $\ddot{\alpha}^a(-T)$. Finally, note that you can write $\Gamma^a_{\mu\nu}(\gamma(-T)) = \Gamma^a_{\mu\nu}(N) + \mathrm{O}(T)$.

(b) Similarly, determine the second worldline $\beta(t)$ of the ball, solving

$$\nabla_{\dot{\beta}}\dot{\beta} = 0, \quad \beta(T) = \gamma(T), \quad \dot{\beta}(T) = \dot{\gamma}(T) - \tilde{w}, \tag{3.4}$$

by Taylor-expanding around $t = T$ to second order.

Hint: You can obtain the solution directly from that for α from part (a) by the substitutions $-T \to T$, $w \to -\tilde{w}$.

(c) For an intersection point P to exist, we need the equation $\alpha(t_P) - \beta(t_P) = 0$ to have a solution t_P. By expanding $t_P = t_P^{(1)}T + t_P^{(2)}T^2 + \mathrm{O}(T^3)$ as a power series in T, expanding $\tilde{w}^a$ analogously, and inserting your approximate worldlines $\alpha(t)$, $\beta(t)$, show that a solution exists if and only if we have

$$\tilde{w}^a = (1 + AT)w^a - 2T\,\Gamma^a_{tb}(N)w^b + \mathrm{O}(T^2) \tag{3.5}$$

for some number A.

Hint: Before inserting the expansions into the equation, determine the leading-order coefficient of $\tilde{w}^a$ by thinking about what $\tilde{w}^a$ looks like for $T = 0$.

(d) Show that parallel transport along γ for time $2T$ maps w to a vector with spatial components

$$w^a - 2T\,\Gamma^a_{tb}(N)w^b + \mathrm{O}(T^2). \tag{3.6}$$

Hint: Solve the parallel transport equation

$$\nabla_{\dot{\gamma}}V(t) = 0, \quad V(-T) = w \tag{3.7}$$

via Taylor expansion around $t = -T$.

Now comparing (3.5) and (3.6), we are finished: parallel transport of spacelike vectors conserves their length (due to $\nabla h = 0$); thus for $\tilde{w}$ to have the same length as w, we need $A = 0$ in (3.5), and $\tilde{w}$ is given by parallel transport. $\quad\triangle$

Interpretation 3.6 Let (M, τ, h) be a Galilei manifold with absolute time and ∇ a torsion-free Galilei connection on it. Consider a timelike curve γ that we interpret as the worldline of an observer. A spacelike vector field w along γ which is parallel with respect to ∇, i.e. satisfies $\nabla_{\dot{\gamma}}w = 0$, has the interpretation of defining a *non-rotating spacelike vector of constant length* from the point of view of the observer moving along γ. $\quad\triangle$

Exercise 3.7 (*Non-rotating vectors in different coordinates*) We consider the $(2+1)$-dimensional Galilei manifold $M = \mathbb{R}^3$ with coordinates $(t, x^1, x^2) = (t, x, y)$, with clock form $\tau = \mathrm{d}t$ and space metric $h = \delta^{ab} \frac{\partial}{\partial x^a} \otimes \frac{\partial}{\partial x^b}$. On this, we consider the torsion-free Galilei connection ∇ whose connection coefficients in the coordinates (t, x, y) vanish, $\Gamma^\rho_{\mu\nu} = 0$, and the unit timelike vector field $v = \frac{\partial}{\partial t} - \omega y \frac{\partial}{\partial x} + \omega x \frac{\partial}{\partial y}$, for some number ω.

(a) Determine the flow line $\gamma_{(x_0, y_0)}(t)$ of v that is determined by the initial condition $\gamma_{(x_0, y_0)}(0) = (0, x_0, y_0)$.

(b) Show that the spacelike vector field $\frac{\partial}{\partial x}$ is non-rotating from the point of view of an observer moving along *any* of the worldlines $\gamma_{(x_0, y_0)}$.
   Hint: In coordinate components, the parallel transport equation $\nabla_{\dot\gamma} w = 0$ becomes $\dot w^\mu(t) + \Gamma^\mu_{\rho\sigma}(\gamma(t)) \, \dot\gamma^\rho(t) w^\sigma(t) = 0$.

(c) We now construct new coordinates $(\tilde t, \tilde x, \tilde y)$ that are 'adapted' to the observer vector field v, in the following way: for any point $p \in M$, there is a unique flow line of v passing through it. We define $(\tilde x(p), \tilde y(p))$ to be the such that this flow line is $\gamma_{(\tilde x(p), \tilde y(p))}$. (This means that the new coordinate system can be imagined to be realised by coordinate labels that are mounted to particles flowing with v.) The time coordinate $\tilde t$ is the same as t. (In these new coordinates, the flow lines of v are the lines of constant $\tilde x, \tilde y$, i.e. $v = \frac{\partial}{\partial \tilde t}$.)
   Determine the form of the non-rotating vector field $\frac{\partial}{\partial x}$ in the new coordinates, i.e. express it in terms of $\frac{\partial}{\partial \tilde x}, \frac{\partial}{\partial \tilde y}$.
   Hint: Using your result from (a), express (x, y) in terms of $(\tilde x, \tilde y)$ in matrix form. From this you can easily obtain $(\tilde x, \tilde y)$ in terms of of (x, y), which allows you to compute $\frac{\partial}{\partial x}$ in terms of the new coordinates.

The components of the non-rotating vector field with respect to the new coordinates seem to 'rotate'. This shows that the coordinates themselves are rotating, in the sense that the lines of constant $\tilde x, \tilde y$—i.e. the flow lines of v—are rotating around each other. $\qquad\qquad\qquad\qquad\qquad\qquad\qquad\qquad\qquad\qquad\qquad\qquad\quad \triangle$

3.2 Kinematics of Timelike Vector Fields

In the following, we will discuss the kinematics of the flow lines of unit timelike vector fields with respect to a Galilei connection.

Definition 3.8 Let (M, τ, h) be a Galilei manifold with absolute time, $\dim M = n + 1$, and ∇ a torsion-free Galilei connection on it. For any unit timelike vector field v, we define its *acceleration* as the timelike part of ∇v,

$$\alpha := \nabla_v v. \tag{3.8a}$$

Further, we decompose the spacelike part $\nabla^\mu v^\nu$ of ∇v into its antisymmetric part, the *twist* of v, given by

$$\omega^{\mu\nu} := \nabla^{[\mu} v^{\nu]}, \tag{3.8b}$$

its trace, the *expansion* of v, given by

$$\theta := h_{\rho\sigma} \nabla^\rho v^\sigma = \nabla_\rho v^\rho, \tag{3.8c}$$

and its symmetric traceless part, the *shear* of v, given by

$$\sigma^{\mu\nu} := \nabla^{(\mu} v^{\nu)} - \frac{1}{n}\theta h^{\mu\nu}. \tag{3.8d}$$

By definition these fields are purely spacelike (or, in the case of θ, a scalar). In terms of them, ∇v can be decomposed as

$$\nabla_\mu v^\nu = \tau_\mu \alpha^\nu + \omega_\mu{}^\nu + \sigma_\mu{}^\nu + \frac{1}{n}\theta P^\nu_\mu, \tag{3.9}$$

and the Newton–Coriolis form of ∇ with respect to v is

$$\Omega_{\mu\nu} = 2\tau_{[\mu}\alpha_{\nu]} + 2\omega_{\mu\nu}. \tag{3.10a}$$

In the following, we will freely use $\underset{v}{h}$ to identify α and ω with differential forms also when *not* using index notation. Using this identification, the Newton–Coriolis form is

$$\Omega = \tau \wedge \alpha + 2\omega. \tag{3.10b}$$

Interpretation 3.9 Let (M, τ, h) be a Galilei manifold with absolute time, and ∇ a torsion-free Galilei connection on it. Let v be a unit timelike vector field, and consider its flow lines. A spacelike[4] vector field η satisfying $\mathcal{L}_v \eta = 0$ is transported along the flow of v, i.e. when evaluated along one fixed flow line γ of v, it may be interpreted as the direction from γ to an 'infinitesimally close' flow line over the course of time; see Fig. 3.2.

Now since ∇ is torsion-free, we have $0 = \mathcal{L}_v \eta = [v, \eta] = \nabla_v \eta - \nabla_\eta v$. This implies

$$v^\mu \nabla_\mu \eta^\nu = \eta^\mu \nabla_\mu v^\nu \overset{(3.9)}{=} \eta^\mu \left(\omega_\mu{}^\nu + \sigma_\mu{}^\nu + \frac{1}{n}\theta P^\nu_\mu \right) = -\omega^\nu{}_\mu \eta^\mu + \sigma^\nu{}_\mu \eta^\mu + \frac{1}{n}\theta \eta^\nu. \tag{3.11a}$$

---

[4] Being everywhere spacelike is compatible with $\mathcal{L}_v \eta = 0$: due to Cartan's magic formula, we have $\mathcal{L}_v \tau = \mathrm{d}(\tau(v)) + \mathrm{d}\tau(v, \cdot) = 0$, which implies $\mathcal{L}_v(\tau(\eta)) = \tau(\mathcal{L}_v \eta) = 0$, i.e. $\tau(\eta)$ is constant along the flow of v.

Fig. 3.2 Two flow lines $\gamma, \widetilde{\gamma}$ of a timelike vector field, and the evolution of the connecting vector η to an 'infinitesimally close' flow line along γ

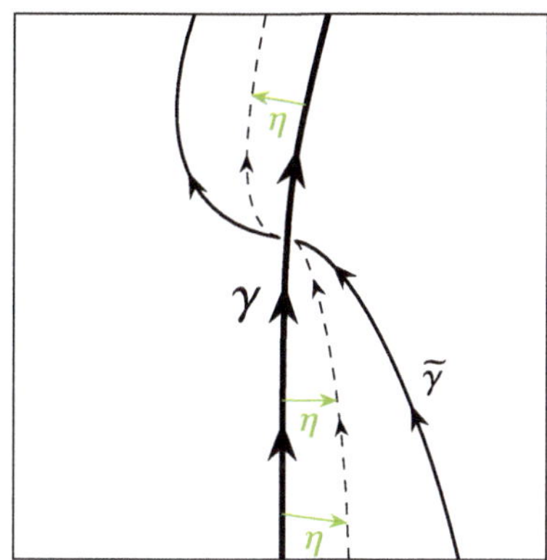

Evaluated along a fixed flow line γ, this gives

$$\nabla_{\dot{\gamma}}\eta = -\omega(\eta) + \sigma(\eta) + \frac{1}{n}\theta\eta, \tag{3.11b}$$

where we treat, at each point, ω and σ as linear maps on the space of spacelike vectors.

This means that the evolution of the 'infinitesimal connecting vector' η along γ, as compared to being non-rotating and of constant length, is at each point in time generated by the linear map $-\omega + \sigma + \frac{1}{n}\theta$ id.[5] Since with respect to the metric $^{(n)}h$ on space, the map ω is antisymmetric and σ is symmetric traceless, ω generates a rotational motion and σ generates a rotation-free volume-preserving motion[6].

Thus ω gives the rate at which 'infinitesimally close' flow lines rotate around each other, σ describes how the shape of a family of points flowing along them is deformed in a volume-preserving way, and θ is the rate at which the volume of an infinitesimal 'ball' of points flowing with v increases isotropically.   △

Exercise 3.10 (*The Raychaudhuri equation*) Let (M, τ, h) be a Galilei manifold with absolute time, and ∇ a torsion-free Galilei connection on it. Let v be a unit timelike vector field, and denote by $\alpha, \omega, \theta, \sigma$ its acceleration, twist, expansion and shear fields with respect to ∇, respectively. Show that the rate of change of the expansion along v is given by the *Raychaudhuri equation*[7]

$$\nabla_v\theta = -\frac{1}{n}\theta^2 - \sigma_{\mu\nu}\sigma^{\mu\nu} + \omega_{\mu\nu}\omega^{\mu\nu} + \nabla_\mu\alpha^\mu - R_{\mu\nu}v^\mu v^\nu, \tag{3.12}$$

where $n = \dim M - 1$, and $R_{\mu\nu}$ are the components of the Ricci tensor of ∇.

---

[5] More explicitly, if e_a is a non-rotating (i.e. parallelly transported) orthonormal frame along γ, and we consider η's components in this frame according to $\eta(\gamma(t)) =: \eta^a(t)e_a(t)$, for the covariant derivative of η along γ we have $\nabla_{\dot{\gamma}(t)}\eta = \dot{\eta}^a(t)e_a(t) + \eta^a(t)\nabla_{\dot{\gamma}}e_a(t) = \dot{\eta}^a(t)e_a(t)$. Using this, the evolution equation (3.11b) becomes $\dot{\eta}^a = -\omega^a{}_b\eta^b + \sigma^a{}_b\eta^b + \frac{1}{n}\theta\eta^a$.

[6] The infinitesimal version of an actual shear in the geometric sense combines this with a rotation.

[7] The exact same equation is true in GR, and is there an important tool for example in the proof of the various singularity theorems.

Hint: Write $\theta = \nabla_\mu v^\mu$, *apply the Ricci identity* $R^\mu{}_{\nu\rho\sigma} X^\nu = 2\nabla_{[\rho}\nabla_{\sigma]}X^\mu$, *and use the decomposition of* ∇v *in terms of the various fields (Eq. (3.9))*. $\triangle$

Proposition 3.11 *Let* (M, τ, h) *be a Galilei manifold with absolute time,* ∇ *a torsion-free Galilei connection on it, and* v *any vector field on* M. *Then the Lie derivative of* h *in direction of* v *is given by* $(\mathcal{L}_v h)^{\mu\nu} = -2\nabla^{(\mu}v^{\nu)}$.

Proof For any one-form α and any vector field X, we have

$$\begin{aligned}
(\mathcal{L}_v \alpha)(X) &= v(\alpha(X)) - \alpha(\mathcal{L}_v X) \\
&= (\nabla_v \alpha)(X) + \alpha(\nabla_v X) - \alpha(\nabla_v X - \nabla_X v) \\
&= (\nabla_v \alpha)(X) + \alpha(\nabla_X v).
\end{aligned} \tag{3.13}$$

In index notation, this reads $(\mathcal{L}_v \alpha)_\rho = v^\mu \nabla_\mu \alpha_\rho + \alpha_\mu \nabla_\rho v^\mu$. This implies that we have

$$h(\mathcal{L}_v \alpha, \beta) = h(\nabla_v \alpha, \beta) + h^{\rho\nu}\alpha_\mu(\nabla_\rho v^\mu)\beta_\nu = h(\nabla_v \alpha, \beta) + (\nabla^\nu v^\mu)\alpha_\mu \beta_\nu \tag{3.14}$$

for one-forms α, β. From this we obtain

$$\begin{aligned}
(\mathcal{L}_v h)(\alpha, \beta) &= v(h(\alpha, \beta)) - h(\mathcal{L}_v \alpha, \beta) - h(\alpha, \mathcal{L}_v \beta) \\
&= h(\nabla_v \alpha, \beta) + h(\alpha, \nabla_v \beta) - h(\mathcal{L}_v \alpha, \beta) - h(\alpha, \mathcal{L}_v \beta) \\
(3.14) \quad &= -(\nabla^\nu v^\mu)\alpha_\mu \beta_\nu - (\nabla^\nu v^\mu)\beta_\mu \alpha_\nu \\
&= -2(\nabla^{(\mu}v^{\nu)})\alpha_\mu \beta_\nu \, .
\end{aligned} \tag{3.15}$$

$\square$

For (unit) timelike v, this matches intuition: $\nabla^{(\mu}v^{\nu)}$ consists of shear and expansion, i.e. encodes (according to Interpretation 3.9) the deformation of spatial geometry along the flow of v, which on the other hand is described by the change of the spatial metric $^{(n)}h$ along the flow of v (the minus sign comes from inverting h to get $^{(n)}h$). Note also that this implies that ∇v is determined by $\mathcal{L}_v h$ and the Newton–Coriolis form Ω.

Definition 3.12 Let (M, τ, h) be a Galilei manifold with absolute time. A unit timelike vector field v is *rigid* if $\mathcal{L}_v h = 0$, i.e. if spatial geometry is constant along its flow.   ✿

Proposition 3.13 *Let* (M, τ, h) *be a Galilei manifold with absolute time, and* v *a unit timelike vector field on it. Then* v *is rigid if and only if the Lie derivative along* v *of the covariant space metric* $\underset{v}{h}$ *vanishes, i.e. we have* $\mathcal{L}_v h = 0 \iff \mathcal{L}_v \underset{v}{h} = 0$.

Proof First, note that due to v being unit timelike, by Cartan's magic formula we have $\mathcal{L}_v \tau = \mathrm{d}(\tau(v)) + \mathrm{d}\tau(v, \cdot) = 0$. The idea is now that, since h and $\underset{v}{h}$ are determined in terms of each other as well as τ and v, and the Lie derivatives of τ and v along v vanish, the Lie derivative of h along v vanishes if and only if that of $\underset{v}{h}$ vanishes.

Explicitly, we decompose $\mathcal{L}_v \underset{v}{h}$ into timelike and spacelike parts, according to

$$(\mathcal{L}_v \underset{v}{h})_{\mu\nu} = (\mathcal{L}_v \underset{v}{h})_{\rho\nu} v^\rho \tau_\mu + (\mathcal{L}_v \underset{v}{h})_{\rho\nu} h^{\rho\sigma} h_{\sigma\mu} \ . \tag{3.16}$$

Taking the Lie derivative of the equation $h_{\rho\nu} v^\rho = 0$, using $\mathcal{L}_v v = 0$ we obtain $0 = (\mathcal{L}_v \underset{v}{h})_{\rho\nu} v^\rho$, i.e. vanishing of the timelike part of (3.16). Similarly, starting from $h_{\rho\nu} h^{\rho\sigma} = P^\sigma_\nu = \delta^\sigma_\nu - v^\sigma \tau_\nu$, by using $\mathcal{L}_v v = 0$ and $\mathcal{L}_v \tau = 0$ we obtain

$$0 = (\mathcal{L}_v \underset{v}{h})_{\rho\nu} h^{\rho\sigma} + h_{\rho\nu} (\mathcal{L}_v h)^{\rho\sigma}. \tag{3.17}$$

This allows us to rewrite the spacelike part of (3.16), giving

$$(\mathcal{L}_v \underset{v}{h})_{\mu\nu} = -(\mathcal{L}_v h)^{\rho\sigma} h_{\rho\nu} h_{\sigma\mu} \ . \tag{3.18}$$

This shows $\mathcal{L}_v h = 0 \implies \mathcal{L}_v \underset{v}{h} = 0$.

Conversely, from $h^{\rho\nu} \tau_\rho = 0$ we obtain $0 = (\mathcal{L}_v h)^{\rho\nu} \tau_\rho$. Using this and (3.17), a timelike-spacelike decomposition of $\mathcal{L}_v h$ yields

$$\begin{aligned} (\mathcal{L}_v h)^{\mu\nu} &= (\mathcal{L}_v h)^{\rho\nu} \tau_\rho v^\mu + (\mathcal{L}_v h)^{\rho\nu} h_{\rho\sigma} h^{\sigma\mu} \\ &= -(\mathcal{L}_v \underset{v}{h})_{\rho\sigma} h^{\rho\nu} h^{\sigma\mu} \ . \end{aligned} \tag{3.19}$$

This shows $\mathcal{L}_v \underset{v}{h} = 0 \implies \mathcal{L}_v h = 0$. $\qquad\qquad\square$

Interpretation 3.14 Using the kinematical decomposition of the covariant derivative of unit timelike vector fields, we can give a kinematical interpretation for the local characterisation of Newtonian connections by Theorem 2.44: a torsion-free Galilei connection is Newtonian if and only if locally around any point there exist timelike flows/unit timelike vector fields that are non-rotating (vanishing twist) and geodesic (vanishing acceleration). Note, however, that in general such a vector field will not be rigid. $\qquad\qquad\triangle$

Proposition 3.15 *Let (M, τ, h) be a Galilei manifold with absolute time, and ∇ a torsion-free Galilei connection on it. For any vector field v with $\nabla^{(\mu} v^{\nu)} = 0$, we have*

$$\nabla^\rho \nabla^\mu v^\nu = -R^{\mu\nu\rho}{}_\sigma v^\sigma. \tag{3.20}$$

Proof Due to vanishing torsion, we have the Ricci identity

$$2\nabla_{[\mu} \nabla_{\nu]} v^\rho = R^\rho{}_{\sigma\mu\nu} v^\sigma. \tag{3.21}$$

Raising μ, ν and cycling indices, we obtain

$$2\nabla^{[\mu}\nabla^{\nu]}v^{\rho} = R^{\rho}{}_{\sigma}{}^{\mu\nu}v^{\sigma}, \tag{3.22a}$$

$$2\nabla^{[\nu}\nabla^{\rho]}v^{\mu} = R^{\mu}{}_{\sigma}{}^{\nu\rho}v^{\sigma}, \tag{3.22b}$$

$$2\nabla^{[\rho}\nabla^{\mu]}v^{\nu} = R^{\nu}{}_{\sigma}{}^{\rho\mu}v^{\sigma}. \tag{3.22c}$$

Adding the last two equations and subtracting the first, we get

$$-2\nabla^{\mu}\nabla^{(\nu}v^{\rho)} + 2\nabla^{\nu}\nabla^{(\rho}v^{\mu)} + 2\nabla^{\rho}\nabla^{[\mu}v^{\nu]} = (-R^{\rho}{}_{\sigma}{}^{\mu\nu} + R^{\mu}{}_{\sigma}{}^{\nu\rho} + R^{\nu}{}_{\sigma}{}^{\rho\mu})v^{\sigma}. \tag{3.23}$$

Due to $\nabla^{(\mu}v^{\nu)} = 0$, the left-hand side of this equation simplifies to $2\nabla^{\rho}\nabla^{\mu}v^{\nu}$. Using the Bianchi identity and the antisymmetries of the curvature tensor, this yields

$$\begin{aligned}
2\nabla^{\rho}\nabla^{\mu}v^{\nu} &= (-R^{\rho}{}_{\sigma}{}^{\mu\nu} + R^{\mu}{}_{\sigma}{}^{\nu\rho} + R^{\nu}{}_{\sigma}{}^{\rho\mu})v^{\sigma} \\
&= (R^{\rho\mu\nu}{}_{\sigma} + R^{\rho\nu}{}_{\sigma}{}^{\mu} - R^{\mu\nu\rho}{}_{\sigma} - R^{\mu\rho}{}_{\sigma}{}^{\nu} - R^{\nu\rho\mu}{}_{\sigma} - R^{\nu\mu}{}_{\sigma}{}^{\rho})v^{\sigma} \\
&= (R^{\rho\mu\nu}{}_{\sigma} + R^{\nu\rho\mu}{}_{\sigma} - R^{\mu\nu\rho}{}_{\sigma} - R^{\rho\mu\nu}{}_{\sigma} - R^{\nu\rho\mu}{}_{\sigma} - R^{\mu\nu\rho}{}_{\sigma})v^{\sigma} \\
&= -2R^{\mu\nu\rho}{}_{\sigma}v^{\sigma}. \tag{3.24}
\end{aligned}$$

$\square$

Corollary 3.16 *Let v be a rigid unit timelike vector field on a Galilei manifold with absolute time with a torsion-free Galilei connection ∇. The twist ω of v satisfies*

$$\nabla^{\rho}\omega^{\mu\nu} = -R^{\mu\nu\rho}{}_{\sigma}v^{\sigma}. \tag{3.25}$$

$\square$

3.3 Recovering Newtonian Gravity

In this section, we will show how to recover the usual coordinate formulation of Newtonian gravity from the geometric framework of Newton–Cartan gravity.

3.3.1 Künzle–Ehlers Recovery

Here, we will step by step show how to recover a slight generalisation of usual Newtonian gravity from Newton–Cartan gravity. We give a formulation of this recovery theorem close to that by Malament [7], but provide a more differential-geometric proof.

In the following, let (M, τ, h) be a Galilei manifold with absolute time, ∇ a torsion-free Galilei connection on it, and v a unit timelike vector field.

Notation 3.17 (i) Let (N, g) be a (pseudo-)Riemannian manifold. For $\beta \in \Omega^k(N)$, we write

$$(\delta\beta)_{\mu_1\ldots\mu_{k-1}} := -\widetilde{\nabla}^\rho \beta_{\rho\mu_1\ldots\mu_{k-1}}\,, \tag{3.26}$$

where $\widetilde{\nabla}$ is the Levi-Civita connection of g. The operator $\delta\colon \Omega^k(N) \to \Omega^{k-1}(N)$ is called the *codifferential*.[8]

(ii) For any spatial leaf Σ of our Galilei manifold (M, τ, h) with absolute time, we will denote by ${}^{(n)}\delta$ the codifferential of the Riemannian manifold $(\Sigma, {}^{(n)}h)$. ✿

Proposition 3.18 *Consider a k-form β and a $(k-1)$-form κ on M that are purely spacelike with respect to v.*

(i) The equation

$$k\nabla^{[\mu_1}\kappa^{\mu_2\ldots\mu_k]} = \beta^{\mu_1\ldots\mu_k} \tag{3.27}$$

is equivalent to

$$\mathrm{d}(\kappa|_\Sigma) = \beta|_\Sigma \text{ on all spatial leaves } \Sigma. \tag{3.28}$$

(ii) The equation

$$-\nabla_\rho\beta^{\rho\mu_1\ldots\mu_{k-1}} = \kappa^{\mu_1\ldots\mu_{k-1}} \tag{3.29}$$

is equivalent to

$${}^{(n)}\delta(\beta|_\Sigma) = \kappa|_\Sigma \text{ on all spatial leaves } \Sigma. \tag{3.30}$$

Proof (i) Since κ is purely spacelike with respect to v, the multivector field with components $\kappa^{\mu_2\ldots\mu_k}$ is purely spacelike (i.e. it may be considered a section not only of $\bigwedge^k TM$, but of $\bigwedge^k \ker\tau$). Hence the expression $\overset{(n)}{\nabla}{}^{\mu_1}\kappa^{\mu_2\ldots\mu_k}$ makes sense, and we have $k\nabla^{[\mu_1}\kappa^{\mu_2\ldots\mu_k]} = k\overset{(n)}{\nabla}{}^{[\mu_1}\kappa^{\mu_2\ldots\mu_k]}$. Restricted to any spatial leaf Σ, these are the components of $\mathrm{d}(\kappa|_\Sigma)$, expressed in terms of the Levi-Civita connection $\overset{(n)}{\nabla}$ of $(\Sigma, {}^{(n)}h)$ (and identifying forms with multivector fields via ${}^{(n)}h$). Thus, restriction of (3.27) to Σ gives (3.28).

(ii) As in the first part, the expression $\overset{(n)}{\nabla}_\rho\beta^{\rho\mu_1\ldots\mu_{k-1}}$ makes sense, and we have $\nabla_\rho\beta^{\rho\mu_1\ldots\mu_{k-1}} = \overset{(n)}{\nabla}_\rho\beta^{\rho\mu_1\ldots\mu_{k-1}}$. Restricted to any spatial leaf Σ, $\overset{(n)}{\nabla}$ is the Levi-Civita connection of $(\Sigma, {}^{(n)}h)$, and thus restriction of (3.29) to Σ gives (3.30). $\square$

---

[8] δ is formally adjoint to the exterior derivative d, i.e. for any k-form α and $(k-1)$-form β with compact support it satisfies $\frac{1}{k!}\int_N \mathrm{vol}_g\, \alpha_{\mu_1\ldots\mu_k}(\mathrm{d}\beta)^{\mu_1\ldots\mu_k} = \frac{1}{(k-1)!}\int_N \mathrm{vol}_g\, (\delta\alpha)_{\mu_1\ldots\mu_{k-1}}\beta^{\mu_1\ldots\mu_{k-1}}$.

It may also be expressed in terms of the exterior derivative d and the Hodge star operator with respect to g, but we will use only (3.26) instead, i.e. 'codifferential is minus covariant divergence'.

In the following, we view the twist ω of v as a two-form and the acceleration α of v as a one-form, both of which are purely spacelike with respect to v.

Construction 3.19 In the following, we are going to rewrite the condition that our connection ∇ be Newtonian as well as the Newton–Cartan field equation in terms of α and ω. Before doing so, we note these conditions' decompositions into spacelike and timelike parts:

The condition that ∇ be Newtonian is equivalent to the Newton–Coriolis form being closed, $\mathrm{d}\Omega = 0$ (Theorem 2.43). Due to antisymmetry, the latter is equivalent to vanishing of both the purely spacelike part of $\mathrm{d}\Omega$, i.e. $(\mathrm{d}\Omega)^{\mu\nu\rho} = 0$, and the mixed timelike-spacelike part of $\mathrm{d}\Omega$, i.e. $v^\mu(\mathrm{d}\Omega)_\mu{}^{\nu\rho} = 0$.

Similarly, the Newton–Cartan field equation (3.1) is satisfied if and only if its purely spacelike part $R^{\mu\nu} = 0$, its mixed spacelike-timelike part $R^\mu{}_\nu v^\nu = 0$, and its purely timelike part $v^\mu v^\nu R_{\mu\nu} = 4\pi G\rho$ hold.   △

Lemma 3.20 *The purely spacelike part of the condition of ∇ being Newtonian, $(\mathrm{d}\Omega)^{\mu\nu\rho} = 0$, holds if and only if on any spatial leaf Σ we have $\mathrm{d}(\omega|_\Sigma) = 0$.*

Proof The spacelike part of the Newtonian condition is equivalent to $(\mathrm{d}\Omega)|_\Sigma = 0$ for all spatial leaves Σ. Now according to (3.10), the twist is one half of the spacelike part of the Newton–Coriolis form, i.e. $\omega|_\Sigma = \frac{1}{2}\,\Omega|_\Sigma$. This implies $\mathrm{d}(\omega|_\Sigma) = \frac{1}{2}\mathrm{d}(\Omega|_\Sigma) = \frac{1}{2}(\mathrm{d}\Omega)|_\Sigma$, since d commutes with pullback and therefore with restriction to submanifolds.   □

Lemma 3.21 *Assume that v is rigid. Then ∇ satisfies the mixed spacelike-timelike part of the Newton–Cartan field equation, $R^\mu{}_\nu v^\nu = 0$, if and only if on any spatial leaf Σ we have $^{(n)}\delta(\omega|_\Sigma) = 0$.*

Proof By Corollary 3.16, we have

$$\nabla_\mu \omega^{\mu\nu} = h_{\rho\mu}\nabla^\rho \omega^{\mu\nu} = -h_{\rho\mu}R^{\mu\nu\rho}{}_\sigma v^\sigma = -R^\nu{}_\sigma v^\sigma \,. \tag{3.31}$$

Thus the considered part of the field equation is equivalent to $\nabla_\mu \omega^{\mu\nu} = 0$, which by Proposition 3.18 (ii) may be invariantly written as $^{(n)}\delta(\omega|_\Sigma) = 0$ for all Σ.   □

Lemma 3.22 *Assume that v is rigid. Then the mixed timelike-spacelike part of the condition of ∇ being Newtonian, $v^\mu(\mathrm{d}\Omega)_\mu{}^{\nu\rho} = 0$, holds if and only if on any spatial leaf Σ acceleration and twist satisfy $\mathrm{d}(\alpha|_\Sigma) = 2(\nabla_v \omega)|_\Sigma$.*

Proof According to (3.10), we have $\Omega = \tau \wedge \alpha + 2\omega$. This implies $\mathrm{d}\Omega = -\tau \wedge \mathrm{d}\alpha + 2\mathrm{d}\omega$, which in components becomes

$$(\mathrm{d}\Omega)_{\mu\nu\rho} = -3\tau_{[\mu}(\mathrm{d}\alpha)_{\nu\rho]} + 2 \cdot 3\partial_{[\mu}\omega_{\nu\rho]} = -6\tau_{[\mu}\partial_\nu\alpha_{\rho]} + 6\partial_{[\mu}\omega_{\nu\rho]} \,. \tag{3.32}$$

Due to ∇ being torsion-free, we may replace the anti-symmetrised partial derivatives by anti-symmetrised covariant derivatives with respect to ∇, obtaining

$$
\begin{aligned}
(\mathrm{d}\Omega)_{\mu\nu\rho} &= -6\tau_{[\mu}\nabla_{\nu}\alpha_{\rho]} + 6\nabla_{[\mu}\omega_{\nu\rho]} \\
&= -2\tau_{\mu}\nabla_{[\nu}\alpha_{\rho]} - 2\tau_{\nu}\nabla_{[\rho}\alpha_{\mu]} - 2\tau_{\rho}\nabla_{[\mu}\alpha_{\nu]} \\
&\quad + 2\nabla_{\mu}\omega_{\nu\rho} + 2\nabla_{\nu}\omega_{\rho\mu} + 2\nabla_{\rho}\omega_{\mu\nu} \; .
\end{aligned}
\tag{3.33}
$$

Then raising ν, ρ and contracting with v^{μ}, we obtain

$$
v^{\mu}(\mathrm{d}\Omega)_{\mu}{}^{\nu\rho} = -2\nabla^{[\nu}\alpha^{\rho]} + 2v^{\mu}\nabla_{\mu}\omega^{\nu\rho} + 2v^{\mu}\nabla^{\nu}\omega^{\rho}{}_{\mu} + 2v^{\mu}\nabla^{\rho}\omega_{\mu}{}^{\nu}.
\tag{3.34}
$$

Due to ω being purely spacelike and v being rigid, using the Leibniz rule we may rewrite the last two terms of (3.34) as

$$
v^{\mu}\nabla^{\nu}\omega^{\rho}{}_{\mu} + v^{\mu}\nabla^{\rho}\omega_{\mu}{}^{\nu} = -\omega^{\rho}{}_{\mu}\nabla^{\nu}v^{\mu} - \omega_{\mu}{}^{\nu}\nabla^{\rho}v^{\mu} = -\omega^{\rho}{}_{\mu}\omega^{\nu\mu} + \omega^{\nu}{}_{\mu}\omega^{\rho\mu} = 0.
\tag{3.35}
$$

Thus, by (3.34) we see that $v^{\mu}(\mathrm{d}\Omega)_{\mu}{}^{\nu\rho} = 0$ is equivalent to

$$
2\nabla^{[\nu}\alpha^{\rho]} = 2v^{\mu}\nabla_{\mu}\omega^{\nu\rho} \; ,
\tag{3.36}
$$

which by Proposition 3.18 (i) may be written as $\mathrm{d}(\alpha|_{\Sigma}) = 2(\nabla_{v}\,\omega)|_{\Sigma}$. $\qquad\square$

Interpretation 3.23 Lemmas 3.20 and 3.22 allow a somewhat 'physical' interpretation of the condition that ∇ be Newtonian.

By Lemma 3.20, the purely spacelike part of the Newtonian condition is equivalent to $\mathrm{d}(\omega|_{\Sigma}) = 0$: the twist of any flow is spatially a closed two-form, i.e. (due to Stokes' theorem) the integral of the twist over the boundary of any three-dimensional spatial volume vanishes. Put differently, this is spatial 'vortex conservation': the amount of twist 'going into' some three-dimensional volume must balance that 'coming out' of it. This is essentially the kinematical version of *Helmholtz's first theorem* on inviscid flows, generalised from three to arbitrary dimensions.[9]

By Lemma 3.22, the mixed timelike-spacelike part of the Newtonian condition is for *rigid* flows equivalent to $\mathrm{d}(\alpha|_{\Sigma}) = 2(\nabla_{v}\,\omega)|_{\Sigma}$: any temporal change of the twist of a rigid flow is due to 'rotational', i.e. non-closed, acceleration. Put differently, rigid flows do not spontaneously change their rate of rotation. Dropping the assumption of v being rigid, the equation generalises to $2v^{\mu}\nabla_{\mu}\omega^{\nu\rho} = 2\nabla^{[\nu}\alpha^{\rho]} + 4\omega^{\mu[\nu}\sigma^{\rho]}{}_{\mu} + 4\frac{\theta}{n}\omega^{\nu\rho}$, showing that if the twist vanishes initially along some flow line and the acceleration is 'non-rotational' (i.e. spatially closed), then it will stay zero along the flow line. Put differently, under non-rotational acceleration, non-rotating flow lines will stay non-rotating over time. This is *Helmholtz's third theorem*. $\qquad\triangle$

---

[9] In the usual coordinate formulation of Newtonian mechanics, this version of Helmholtz's first theorem is in fact a provable statement, whose proof is very direct: it is just an application of the general fact that the divergence of a curl vanishes (or, put differently, that the exterior derivative satisfies $\mathrm{d}^{2} = 0$). Here in the geometric formulation in terms of Galilei connections, however, it becomes a kinematical assumption about ∇.

Now, in addition to ∇ we consider the special Galilei connection $\overset{v}{\nabla}$ with respect to v, i.e. the unique torsion-free Galilei connection that has vanishing Newton–Coriolis form with respect to v. According to the classification theorem (Theorem 2.29), we have

$$\Gamma^{\rho}_{\mu\nu} = \overset{v}{\Gamma}{}^{\rho}_{\mu\nu} + \tau_{(\mu}\Omega_{\nu)}{}^{\rho}. \tag{3.37}$$

Corollary 3.24 *Assume that v is rigid. Then the mixed timelike-spacelike part of the condition of ∇ being Newtonian is equivalent to* $\mathrm{d}(\alpha|_{\Sigma}) = 2(\overset{v}{\nabla}_v \omega)|_{\Sigma}$ *on any spatial leaf Σ.*

Proof A direct calculation using (3.37) and (3.10) shows that $\nabla_v \omega = \overset{v}{\nabla}_v \omega$, so this is a restatement of Lemma 3.22. $\qquad\square$

Lemma 3.25 *If v is rigid, then $\overset{v}{\nabla} v = 0$.*

Proof Since v is unit timelike, the index ν of $\overset{v}{\nabla}_\mu v^\nu$ is spacelike. Since v is rigid, we have $\overset{v}{\nabla}{}^{(\mu}v^{\nu)} = 0$; since the Newton–Coriolis form of $\overset{v}{\nabla}$ with respect to v vanishes, we also have $\overset{v}{\nabla}{}^{[\mu}v^{\nu]} = 0$ and $\overset{v}{\nabla}_v v = 0$. This means that both the purely spacelike and the mixed timelike-spacelike part of ∇v vanish, i.e. $\overset{v}{\nabla} v = 0$. $\qquad\square$

Lemma 3.26 *Let v be rigid, and (M, τ, h) spatially flat. Then $\overset{v}{\nabla}$ is flat.*

Proof Fix any spatial leaf Σ. Since it is flat as a Riemannian manifold (spatial flatness), there exists (locally) a parallel frame of vector fields on Σ, i.e. vector fields $\tilde{e}_a \in \Gamma(T\Sigma)$, $a \in \{1, \ldots, n\}$ which satisfy $\overset{(n)}{\nabla}_{\tilde{e}_a} \tilde{e}_b = 0$ and pointwise are a basis of $T_p\Sigma$.

We extend these vector fields along the flow of v to obtain vector fields e_a defined on a neighbourhood of Σ in M, i.e.

$$e_a|_{\Sigma} = \tilde{e}_a \, , \quad 0 = \mathcal{L}_v e_a = [v, e_a]. \tag{3.38}$$

Due to vanishing torsion and $\overset{v}{\nabla} v = 0$ (Lemma 3.25), we have

$$0 = [v, e_a] = \overset{v}{\nabla}_v e_a - \overset{v}{\nabla}_{e_a} v = \overset{v}{\nabla}_v e_a \, . \tag{3.39}$$

Furthermore, since on each spatial leaf $\overset{(n)}{\nabla}$ is the Levi-Civita connection of the induced spatial metric, the covariant derivative $\overset{(n)}{\nabla}_{e_a} e_b$ can be expressed purely in terms of h, e_a and e_b by means of tensor products, contraction, and (exterior) differentiation of scalar functions. The Lie derivative commutes with these operations, so due

to $\mathcal{L}_v \overset{v}{h} = 0$ (Proposition 3.13), $\mathcal{L}_v e_a = 0$, and $\mathcal{L}_v e_b = 0$, we have $\mathcal{L}_v(\overset{(n)}{\nabla}_{e_a} e_b) = 0$.
Together with $(\overset{(n)}{\nabla}_{e_a} e_b)|_\Sigma = \overset{(n)}{\nabla}_{\tilde{e}_a} \tilde{e}_b = 0$, this implies

$$0 = \overset{(n)}{\nabla}_{e_a} e_b = \overset{v}{\nabla}_{e_a} e_b \ . \tag{3.40}$$

Combined, we thus have a local frame $\{v, e_a\}$ that satisfies

$$\overset{v}{\nabla} v = 0, \quad \overset{v}{\nabla} e_a = 0, \tag{3.41}$$

i.e. that is parallel with respect to $\overset{v}{\nabla}$. This means that $\overset{v}{\nabla}$ is flat. $\qquad\square$

Lemma 3.27 *Assume that v is rigid and (M, τ, h) is spatially flat. Then ∇ satisfies the purely timelike part of the Newton–Cartan field equation, $v^\mu v^\nu R_{\mu\nu} = 4\pi G \rho$, if and only if on any spatial leaf Σ we have*

$$- {}^{(n)}\delta(\alpha|_\Sigma) = 4\pi G \rho - \omega_{\mu\nu}\omega^{\mu\nu}. \tag{3.42}$$

Proof Writing $\Gamma^\rho_{\mu\nu} = \overset{v}{\Gamma}{}^\rho_{\mu\nu} + S^\rho{}_{\mu\nu}$, the curvature tensor of ∇ can be expressed in terms of the curvature tensor $\overset{v}{R}$ of $\overset{v}{\nabla}$ as

$$R^\mu{}_{\nu\rho\sigma} = \overset{v}{R}{}^\mu{}_{\nu\rho\sigma} + \overset{v}{\nabla}_\rho S^\mu{}_{\sigma\nu} - \overset{v}{\nabla}_\sigma S^\mu{}_{\rho\nu} + S^\mu{}_{\rho\lambda} S^\lambda{}_{\sigma\nu} - S^\mu{}_{\sigma\lambda} S^\lambda{}_{\rho\nu} \ . \tag{3.43}$$

Hence, due to flatness of $\overset{v}{\nabla}$ (Lemma 3.26) we have

$$\begin{aligned}
v^\nu v^\sigma R_{\nu\sigma} &= v^\nu v^\sigma R^\mu{}_{\nu\mu\sigma} \\
&= v^\nu v^\sigma (\overset{v}{\nabla}_\mu S^\mu{}_{\sigma\nu} - \overset{v}{\nabla}_\sigma S^\mu{}_{\mu\nu} + S^\mu{}_{\mu\lambda} S^\lambda{}_{\sigma\nu} - S^\mu{}_{\sigma\lambda} S^\lambda{}_{\mu\nu}). \tag{3.44}
\end{aligned}$$

Equation (3.37) together with $\Omega = \tau \wedge \alpha + 2\omega$ (Eq. (3.10)) gives

$$S^\rho{}_{\mu\nu} = \tau_{(\mu}\Omega_{\nu)}{}^\rho = \tau_\mu \tau_\nu \alpha^\rho + 2\tau_{(\mu}\omega_{\nu)}{}^\rho, \tag{3.45}$$

and in particular $S^\mu{}_{\mu\nu} = 0$. Using these, (3.44) becomes

$$\begin{aligned}
v^\nu v^\sigma R_{\nu\sigma} &= v^\nu v^\sigma \big(\tau_\sigma \tau_\nu \overset{v}{\nabla}_\mu \alpha^\mu + 2\tau_{(\sigma} \overset{v}{\nabla}_{|\mu|}\omega_{\nu)}{}^\mu \\
&\quad - (\tau_\sigma \tau_\lambda \alpha^\mu + 2\tau_{(\sigma}\omega_{\lambda)}{}^\mu)(\tau_\mu \tau_\nu \alpha^\lambda + 2\tau_{(\mu}\omega_{\nu)}{}^\lambda)\big) \\
&= \overset{v}{\nabla}_\mu \alpha^\mu + 2v^\nu \overset{v}{\nabla}_\mu \omega_\nu{}^\mu - \omega_\lambda{}^\mu \omega_\mu{}^\lambda, \tag{3.46}
\end{aligned}$$

where we have used that α and ω are purely spacelike. Due to $\overset{v}{\nabla} v = 0$ (Lemma 3.25), we have $v^\nu \overset{v}{\nabla}_\mu \omega_\nu{}^\mu = \overset{v}{\nabla}_\mu(v^\nu \omega_\nu{}^\mu) = 0$. Thus (3.46) is equivalent to

$$v^\mu v^\nu R_{\mu\nu} = \overset{v}{\nabla}_\mu \alpha^\mu + \omega_{\mu\nu}\omega^{\mu\nu}, \tag{3.47}$$

which together with Proposition 3.18 (ii) finishes the proof. $\qquad\square$

Combining everything, we arrive at the final result.

Theorem 3.28 *(Künzle–Ehlers recovery theorem) Let (M, τ, h, ∇) be a Newtonian manifold that satisfies the Newton–Cartan field equation and is spatially flat. Let v be a rigid unit timelike vector field on (M, τ, h), let α and ω be its acceleration and twist with respect to ∇, and denote by $\overset{v}{\nabla}$ the special Galilei connection with respect to v. Then*

(i) $\overset{v}{\nabla}v = 0$,

(ii) $\overset{v}{\nabla}$ is flat,

(iii) a timelike curve γ parametrised by absolute time is a geodesic of ∇ if and only if

$$(\overset{v}{\nabla}_{\dot{\gamma}}\dot{\gamma})^{\nu} = -\alpha^{\nu} + 2\omega^{\nu}{}_{\mu}\dot{\gamma}^{\mu} \tag{3.48}$$

along γ, and

(iv) on any spatial leaf Σ, we have the 'recovered field equations'

$$\mathrm{d}(\omega|_{\Sigma}) = 0, \qquad\qquad {}^{(n)}\delta(\omega|_{\Sigma}) = 0, \tag{3.49a}$$

$$\mathrm{d}(\alpha|_{\Sigma}) = 2(\overset{v}{\nabla}_{v}\,\omega)|_{\Sigma}\,, \qquad -{}^{(n)}\delta(\alpha|_{\Sigma}) = 4\pi G\rho - \omega_{\mu\nu}\omega^{\mu\nu}, \tag{3.49b}$$

where ρ is the mass density.

Proof This is a combination of Lemmas 3.20, 3.21, 3.25, 3.26 and 3.27 and Corollary 3.24. The only remaining statement that we have to prove is (3.48), which follows easily from (3.37) and (3.10): for any curve γ with $\tau(\dot{\gamma}) = 1$, these give

$$(\nabla_{\dot{\gamma}}\dot{\gamma})^{\nu} = (\overset{v}{\nabla}_{\dot{\gamma}}\dot{\gamma})^{\nu} + \tau_{(\mu}\Omega_{\lambda)}{}^{\nu}\dot{\gamma}^{\mu}\dot{\gamma}^{\lambda} = (\overset{v}{\nabla}_{\dot{\gamma}}\dot{\gamma})^{\nu} + (\tau_{\mu}\tau_{\lambda}\alpha^{\nu} + 2\tau_{(\mu}\omega_{\lambda)}{}^{\nu})\dot{\gamma}^{\mu}\dot{\gamma}^{\lambda}$$

$$= (\overset{v}{\nabla}_{\dot{\gamma}}\dot{\gamma})^{\nu} + \alpha^{\nu} - 2\omega^{\nu}{}_{\lambda}\dot{\gamma}^{\lambda}. \tag{3.50}$$

$\square$

Construction 3.29 *(Recovery in adapted coordinates)*   Suppose that our Galilei manifold (M, τ, h) is spatially flat. On one spatial leaf Σ, we (locally) introduce orthonormal coordinates $(\tilde{x}^{a})$, $a \in \{1, \dots, n\}$, i.e. coordinates in which the metric takes the form ${}^{(n)}h_{ab} = \delta_{ab}$ (these exist since $(\Sigma, {}^{(n)}h)$ is a flat Riemannian manifold). Now using our unit timelike vector field v, we can extend the coordinates to a neighbourhood of Σ in M: starting at a point on Σ with coordinates (x^{a}) and following the flow of v for time parameter t, the resulting point in M gets coordinates (t, x^{a}). Put differently, the coordinates are defined such that the flow lines of v are lines of constant spatial coordinates (x^{a}) and varying t, i.e.

$$v = \partial_{t} \text{ with respect to } (t, x^{a}); \tag{3.51}$$

and the (x^{a}) can be imagined to be realised by coordinate labels that are mounted to particles flowing with v, see Fig. 3.3. Since v is unit timelike, its flow time coordinate

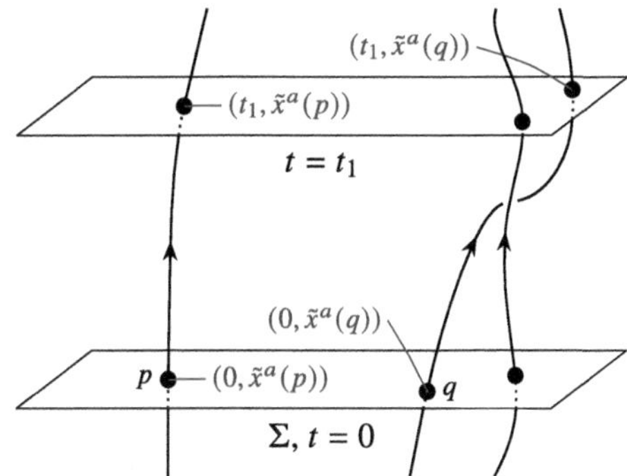

Fig. 3.3 The definition of coordinates adapted to a timelike vector field

t is absolute time, i.e. it satisfies $dt = \tau$. Note that in index notation, we write components corresponding to the t coordinate with an index t.

Since the spacelike coordinate vector fields $e_a := \partial_a$ are orthonormal on Σ, and v is unit timelike, at each point $p \in \Sigma$ the coordinate vector fields (v, e_a) form a Galilei basis. This means in particular that $h = \delta^{ab} e_a \otimes e_b$ on Σ, i.e.

$$h^{t\mu} = h^{\mu t} = 0, \; h^{ab} = \delta^{ab} \text{ in the coordinates } (t, x^a) \text{ on } \Sigma. \qquad (3.52)$$

Now since $v = \partial_t$ is a coordinate vector field, its rigidness, $\mathcal{L}_v h = 0$, means that the coordinate components $h^{\mu\nu}$ are constant with respect to v, i.e. $\partial_t h^{\mu\nu} = 0$. Hence we have

$$h^{t\mu} = h^{\mu t} = 0, \; h^{ab} = \delta^{ab} \text{ in the coordinates } (t, x^a) \qquad (3.53)$$

in the whole domain of definition of the coordinates, i.e. $h = \delta^{ab} e_a \otimes e_b$. Together with v being unit timelike, this implies that the coordinate vector fields (v, e_a) form a Galilei basis at each point of their domain. Thus, the covariant space metric with respect to v is $\underset{v}{h} = \delta_{ab} dx^a \otimes dx^b$, i.e. we have

$$h_{t\mu} = h_{\mu t} = 0, \; h_{ab} = \delta_{ab} \text{ in the coordinates } (t, x^a). \qquad (3.54)$$

Since the coordinate components $h_{\mu\nu}$ and $\tau_\mu = \delta^t_\mu$ are constant, according to (2.37) the special connection $\overset{v}{\nabla}$ therefore has components

$$\overset{v}{\Gamma}{}^\rho_{\mu\nu} = 0 \text{ in the coordinates } (t, x^a). \qquad (3.55)$$

Expressed in those adapted coordinates, the test particle equation of motion (3.48) therefore takes the form

$$\ddot{\gamma}^a = -\alpha^a + 2\omega^a{}_b \dot{\gamma}^b. \qquad (3.56)$$

This means that in those coordinates $-\alpha$ plays the role of the Newtonian gravitational field (gravitational acceleration), while the local rate $-\omega$ of rotation of the spatial frame (e_a) gives rise to a Coriolis-force term (in the three-dimensional case, writing $-\omega_{ab} =: \varepsilon_{abc}\omega^c$, the term becomes the familiar $2\omega^a{}_b\dot{\gamma}^b = -2(\vec{\omega} \times \dot{\vec{\gamma}})^a)$. This is the reason for $\Omega = \tau \wedge \alpha + 2\omega$ being called the Newton–Coriolis form.

The recovered field equations (3.49) become

$$\partial_{[a}\omega_{bc]} = 0, \qquad\qquad \partial_a\omega^{ab} = 0, \qquad\qquad (3.57a)$$

$$\partial_{[a}\alpha_{b]} = \partial_t\omega_{ab}, \qquad\qquad \partial_a\alpha^a = 4\pi G\rho - \omega_{ab}\omega^{ab}, \qquad\qquad (3.57b)$$

i.e. on each spatial leaf the twist field ω has to satisfy the constraints (3.57a), and by (3.57b) mass density and twist together determine the gravitational field $-\alpha$ (up to homogeneous solutions of (3.57b), i.e. harmonic one-forms).

We thus have recovered the usual formulation of Newtonian gravity (in n-dimensional space) in Cartesian coordinates with respect to potentially rotating reference frames. $\qquad\qquad\qquad\qquad\qquad\qquad\qquad\qquad\qquad\qquad\qquad\qquad\qquad\triangle$

The formulation of the recovery theorem in terms of the flat special connection $\overset{v}{\nabla}$ may be seen as a coordinate-free method to state what the equations look like in coordinates adapted to the 'observer vector field' v.

Exercise 3.30 (*Recovery under change of frame*)   Let v and $\tilde{v}$ be rigid unit timelike vector fields on our Galilei manifold (M, τ, h), and let w be the Milne boost field relating v and $\tilde{v}$, i.e. $w := v - \tilde{v}$. In this exercise, we are going to show how the acceleration and twist fields appearing in the recovery theorem behave under this change of vector field.

(a) Show that, given v is rigid, the condition that $\tilde{v}$ be rigid as well is equivalent to

$$\nabla^{(\mu}w^{\nu)} = 0. \qquad\qquad (3.58)$$

(b) Show that the acceleration and twist of $\tilde{v}$ are given in terms of those of v by

$$\tilde{\alpha}^\mu = \alpha^\mu - (\nabla_v w)^\mu + \omega^\mu{}_\nu w^\nu + (\nabla_w w)^\mu , \qquad \tilde{\omega}^{\mu\nu} = \omega^{\mu\nu} - \nabla^\mu w^\nu . \qquad (3.59)$$

Hint: This is a direct computation, using that both v and $\tilde{v}$ are rigid and w is spacelike.

(c) Show that in terms of the special Galilei connection $\overset{v}{\nabla}$, the equation for the acceleration takes the form

$$\tilde{\alpha}^\mu = \alpha^\mu - (\overset{v}{\nabla}_v w)^\mu + 2\omega^\mu{}_\nu w^\nu + (\overset{v}{\nabla}_w w)^\mu . \qquad\qquad (3.60)$$

Hint: Since both ∇ and $\overset{v}{\nabla}$ are torsion-free Galilei connections, on the spatial leaves of M both induce the Levi-Civita connection. With w being spacelike, we therefore have $\nabla_w w = \overset{v}{\nabla}_w w$. Hence you only need to compute $\nabla_v w$ in terms of $\overset{v}{\nabla}_v w$. $\qquad\qquad\qquad\qquad\qquad\qquad\qquad\qquad\qquad\qquad\qquad\qquad\qquad\triangle$

The converse of the recovery theorem is true as well:

Theorem 3.31 *(Geometrisation theorem) Let (M, τ, h) be a Galilei manifold with absolute time that is spatially flat, and v a rigid unit timelike vector field on it. We know that the associated special Galilei connection $\overset{v}{\nabla}$ is flat.*

Given any one-form α and two-form ω that are purely spacelike with respect to v, consider the torsion-free Galilei connection ∇ that has Newton–Coriolis form $\Omega :=\tau \wedge \alpha + 2\omega$ with respect to v. By construction, α and ω are then the acceleration and twist of v with respect to ∇. Furthermore,

(i) *a timelike curve γ parametrised by absolute time satisfies the standard Newtonian gravity equation of motion (3.48) with respect to $\overset{v}{\nabla}$, with $-\alpha$ as the Newtonian gravitational acceleration and $-\omega$ the local frame rotation rate, if and only if it is a geodesic of ∇, and*

(ii) *if α and ω satisfy the recovered field equations (3.49) on every spatial leaf Σ, the resulting ∇ will be Newtonian (i.e. $\mathrm{d}\Omega = 0$) and satisfy the Newton–Cartan field equation.*

Proof The equivalence of (3.48) and the geodesic equation with respect to ∇ was shown in the proof of the recovery theorem (Theorem 3.28). As for the recovery theorem, most of (ii) is a combination of (the converse directions of) Lemmas 3.20, 3.21, 3.27 and 3.24. The only remaining statement in need of proof is the purely spacelike part of the Newton–Cartan field equation, $R^{\mu\nu} = 0$, or equivalently vanishing of the Ricci tensor on spacelike vectors. However, (M, τ, h) being spatially flat means that all spatial leaves are flat as Riemannian manifolds. In particular they are Ricci-flat, which by Construction 2.24 means that the (spacetime) Ricci tensor vanishes on spacelike vectors. $\qquad\square$

Construction 3.32 We can even start one step further back: given a solution α, ω to the equations (3.57) on (some portion of) $M = \mathbb{R}^{n+1}$ with coordinates (t, x^a), we can define $\tau = \mathrm{d}t$, $h := \delta^{ab}\partial_a \otimes \partial_b$, and $v := \partial_t$, and then apply Theorem 3.31 to construct a spacetime satisfying the axioms for Newton–Cartan gravity. The (coordinate) test particle equation of motion (3.56) translates to the geodesic equation for the Newtonian connection of Newton–Cartan gravity.

This construction is why Theorem 3.31 is called a 'geometrisation theorem': it allows 'geometrisation' of a solution of standard Newtonian gravity (allowing for rotating reference frames) into a Newton–Cartan spacetime. $\qquad\triangle$

Note that the recovery theorem depends crucially on the rigidity of the 'observer vector field' v with respect to which we have performed the recovery (i.e. which we use, in the coordinate formulation of Construction 3.29, to identify the spatial coordinates on different spatial leaves). If there were no rigid vector fields, we would not be able to recover Newtonian gravity. But fortunately spatial flatness ensures that we may construct rigid vector fields:

Proposition 3.33 *Let (M, τ, h) be a Galilei manifold with absolute time that is spatially flat. Then for any timelike worldline γ one can construct a rigid timelike vector field v (in a neighbourhood of γ) that along γ agrees with $\dot{\gamma}$.*

Proof Since the spatial leaves are flat, on each of them we can introduce orthonormal coordinates for the induced metric. In particular, we can introduce coordinates (t, x^a) in a neighbourhood of γ in M such that t is absolute time, the x^a are orthonormal coordinates on each spatial leaf, and $x^a(\gamma(t)) = 0$ for all t.[10] The x^a being orthonormal with respect to the induced metric $^{(n)}h$ means that in the coordinates (t, x^a), we have $^{(n)}h_{ab} = \delta_{ab}$. Together with $\tau = \mathrm{d}t$ this implies that the coordinate vector fields (∂_t, ∂_a) form a Galilei basis at each point where the coordinates are defined. This implies that

$$h^{t\mu} = h^{\mu t} = 0, \ h^{ab} = \delta^{ab} \text{ in the coordinates } (t, x^a). \tag{3.61}$$

Now we define

$$v := \partial_t \text{ in the coordinates } (t, x^a), \tag{3.62}$$

i.e. v points in the direction of constant spatial coordinates x^a. By construction, v agrees with $\dot{\gamma}$ along γ.

Furthermore, the components of the Lie derivative $\mathcal{L}_v h$ in our coordinates are given by $(\mathcal{L}_v h)^{\mu\nu} = (\mathcal{L}_{\partial_t} h)^{\mu\nu} = \partial_t h^{\mu\nu} = 0$, since v is a coordinate vector field and the coordinate components $h^{\mu\nu}$ are constant. This shows $\mathcal{L}_v h = 0$, i.e. rigidity of v. $\qquad\square$

3.3.2 Trautman's Condition of Absolute Rotation

In the Künzle–Ehlers recovery theorem (Theorem 3.28), the twist field ω of the 'observer vector field' v with respect to which we perform the recovery appears. Introducing adapted coordinates as in Construction 3.29, this corresponds to a Coriolis-force term in the 'standard Newtonian gravity' test particle equation of motion (3.56), i.e. to rotation of the reference frame defined by observers moving with the flow of v. In the case $\omega = 0$, i.e. for *twist-free* observer vector fields v, we would arrive at the standard formulation of Newtonian gravity in inertial reference frames.

However, if we *had* a twist-free rigid field v, by Corollary 3.16 this would imply $R^{\mu\nu\rho}{}_\sigma v^\sigma = -\nabla^\rho \omega^{\mu\nu} = 0$, which in general need not be true: in the general case of

---

[10] Explicitly, such coordinates may be constructed by taking a smooth orthonormal frame of spacelike vector fields e_a defined along γ, and on each leaf Σ at time t introducing normal coordinates around $\gamma(t)$ with respect to the basis $\{e_a(t)\}$ of $T_{\gamma(t)}\Sigma$.

In the following construction of v, the e_a become the 'connecting vectors' from γ to 'infinitesimally close' flow lines of v, so by choosing the e_a to be non-rotating (i.e. parallelly transported) along γ, we could arrange for the twist of v to vanish along γ.

Newton–Cartan gravity, *twist-free rigid vector fields need not exist*, due to a 'curvature obstruction'. In the following, we will discuss an additional curvature condition which is demanded in the formulation of Newton–Cartan gravity by Trautman [13], and, as we will see, guarantees the existence of twist-free rigid vector fields.

Construction 3.34 Consider a spatially flat Galilei manifold with absolute time with a torsion-free Galilei connection ∇. Due to spatial flatness, the connection's curvature tensor satisfies $R^{\mu\nu\rho}{}_\sigma (v^\sigma - \tilde{v}^\sigma) = 0$ for any two unit timelike vector fields v and $\tilde{v}$. For rigid fields, by Corollary 3.16 this implies that

$$\nabla^\rho \omega^{\mu\nu} = \nabla^\rho \tilde{\omega}^{\mu\nu}, \tag{3.63}$$

i.e. the spatial derivative of the twist is the same for all rigid unit timelike vector fields.

This generalises our above observation on the generic non-existence of twist-free rigid vector fields: if the derivative (3.63), which is given by $R^{\mu\nu\rho}{}_\sigma v^\sigma$, is non-vanishing, then the twist of any rigid unit timelike vector field is spatially varying (in the same way). This means that the spacetime modelled by our manifold does not allow for an absolute notion of rotation for rigid motion: even if a rigid flow is locally non-rotating (vanishing twist) at one point on some spatial leaf, at other points it will be rotating (non-vanishing twist).

If instead the derivative (3.63) vanishes, the twist of any rigid unit timelike vector field is spatially constant. In this case, we do have an absolute notion of rotation for rigid motion: on any spatial leaf a given rigid flow has a constant twist, i.e. local rate of rotation/angular velocity of flow lines around each other. $\triangle$

Proposition 3.35 *Consider a spatially flat Galilei manifold with absolute time with a torsion-free Galilei connection ∇. The following statements are equivalent:*

(i) The connection's curvature tensor satisfies

$$R^{\mu\nu}{}_{\rho\sigma} = 0. \tag{3.64}$$

(ii) For all timelike vectors ξ, we have

$$R^{\mu\nu\rho}{}_\sigma \xi^\sigma = 0. \tag{3.65}$$

(iii) Given any rigid unit timelike vector field, its twist ω is spatially constant, i.e.

$$\nabla^\rho \omega^{\mu\nu} = 0. \tag{3.66}$$

(iv) There is one *rigid unit timelike vector field whose twist is spatially constant.*

Proof Equivalence of (i) and (ii) follows from spatial flatness and antisymmetry of the curvature tensor in its last two indices.

Given (ii), we directly obtain (iii) by Corollary 3.16. Conversely, given any time-like vector $\xi \in T_p M$, which without loss of generality we may assume unit timelike, we can extend it to some timelike curve γ in a neighbourhood of p, which by Proposition 3.33 can in turn be extended to a rigid vector field v, which then satisfies $v|_p = \xi$. By Corollary 3.16, we then have $R^{\mu\nu\rho}{}_\sigma \xi^\sigma = (R^{\mu\nu\rho}{}_\sigma v^\sigma)|_p = -(\nabla^\rho \omega^{\mu\nu})|_p$, and thus (iii) implies (ii).

Finally, as discussed in Construction 3.34, spatial flatness implies that the spatial derivative of the twist of any two rigid vector fields is equal, implying equivalence of (iii) and (iv). $\qquad\square$

Definition 3.36 A spatially flat Galilei manifold with absolute time with a torsion-free Galilei connection is said to have *absolute rotation* if it satisfies the equivalent conditions from Proposition 3.35. ❀

The name 'absolute rotation' is of course due to the fact that, as discussed in Construction 3.34, in a spacetime satisfying these conditions we have an absolute notion of rotation for rigid motions. This kinematical interpretation is most obvious in the 'spatially constant twist' formulation (iii) of the absolute rotation condition; however the formulation (i) as the curvature condition (3.64) is easiest to state (since it involves no additional fields). The original formulation of the condition by Trautman [13] was the equivalent form $\tau_{[\kappa} R^\mu{}_{\nu]\rho\sigma} = 0$ of the curvature condition (3.64); the formulation (3.64) itself is due to Dixon [1].[11]

Remark 3.37 The possibility of motions that are rigid but whose angular velocity varies throughout space might seem counter-intuitive, and so one might be tempted to discard Newton–Cartan spacetimes not having absolute rotation as 'unphysical'— after all, in usual Newtonian mechanics, rotation *is* absolute. However, when considering Newton–Cartan gravity as a limit of GR (see Sect. 3.4), the absolute rotation condition is not automatically satisfied; i.e. there are general-relativistic spacetimes with 'Newtonian' limits in which rotation is *not* absolute. In Exercise 3.55 we will see an explicit example for this. $\qquad\triangle$

Exercise 3.38 (*Non-absolute rotation*) In this exercise, we construct a Newton–Cartan spacetime—although in a rather *ad hoc* fashion—that fails to satisfy the absolute rotation condition, i.e. in which rigid vector fields have spatially non-constant twist.

---

[11] Trautman gave the interpretation of the curvature condition as assuring that (with a specific choice of, in our language, observer vector field v) the Newtonian gravitational field be irrotational, i.e. the spatial gradient of a potential function. However, the interpretation in terms of the existence of an absolute notion of rotation for rigid motions was not explicitly given in reference [13]. To my (the author's) knowledge, it first appears in explicit form in Ehlers' 1981 article on frame theory [2,3].

By the geometrisation construction (Construction 3.32), it suffices to consider solutions to the field equations for twist and acceleration in adapted coordinates, (3.57). We want to find a solution to these equations where the ω^{ab} are spatially non-constant.

(a) Explain why we need to work in $n \geq 3$ spatial dimensions to find solutions with spatially varying ω.

(b) In $n = 3$ dimensions, we may write any spatial two-form as $\omega_{ab} = -\varepsilon_{abc}\omega^c$, where ε_{abc} is the usual three-dimensional totally antisymmetric symbol[12]. Show that then the field equations (3.57a) take the form

$$\vec{\nabla} \cdot \vec{\omega} = 0, \quad \vec{\nabla} \times \vec{\omega} = 0, \tag{3.67}$$

where we use the standard notation for divergence and curl in vector calculus on three-dimensional Euclidean space.

(c) Find a non-constant vector field $\vec{\omega}$ on $\mathbb{R}^3$ that satisfies (3.67).
Hint: You can simply guess a solution: take $\omega^x(x, y, z)$ to be the easiest non-constant function you can think of, and work from there.

(d) Rewrite (3.57b) in terms of $\vec{\omega}$ and usual vector calculus operations. Insert your $\vec{\omega}$ from (c) and find a solution $\vec{\alpha}$ for the case $\rho = 0$. △

If we have absolute rotation, we can construct twist-free rigid vector fields:

Proposition 3.39 *Let (M, τ, h, ∇) be a spatially flat Galilei manifold with absolute time with a torsion-free Galilei connection that has absolute rotation. Let v be a rigid unit timelike vector field, and γ one fixed flow line of v. Then there is a unique twist-free rigid timelike field $\tilde{v}$ that along γ agrees with v.*

Proof The idea of the proof is that due to v 'rotating everywhere with the same speed', we may 'counter-rotate'.

Due to rigidity, we have $\nabla^\mu v^\nu = \omega^{\mu\nu}$, and $\tilde{v}$ being rigid and twist-free would mean $\nabla^\mu \tilde{v}^\nu = 0$. This is true if and only if the spacelike difference vector field $w := v - \tilde{v}$ satisfies

$$\overset{(n)}{\nabla}{}^\mu w^\nu = \omega^{\mu\nu}, \quad w|_\gamma = 0. \tag{3.68}$$

This is an independent equation for w on each spatial leaf Σ. Due to spatial flatness, we may introduce orthonormal coordinates (x^a) on Σ such that $x^a(\gamma(t)) = 0$ for the corresponding value of t. In these coordinates, the equation we need to solve becomes

$$\partial_a w^b = \omega_a{}^b, \quad w^a(0) = 0, \tag{3.69}$$

---

[12] Geometrically, it may be understood as the components of the natural volume form on Euclidean space in orthonormal coordinates.

where the components $\omega_a{}^b$ are constant ($\omega|_\Sigma$ is covariantly constant, i.e. its components in orthonormal coordinates are constant). This system has the unique solution

$$w^a(x) = \omega_b{}^a x^b. \tag{3.70}$$

$\square$

Theorem 3.40 (*Trautman recovery theorem*)   *Let (M, τ, h, ∇) be a Newtonian manifold that satisfies the Newton–Cartan field equation, is spatially flat, and has absolute rotation. Let v be a twist-free rigid unit timelike vector field on (M, τ, h, ∇), and denote by $\overset{v}{\nabla}$ the special Galilei connection with respect to v. Then*

(i) locally there exists a function ϕ on M, unique up to addition of a constant on each spatial leaf, such that a timelike curve γ parametrised by absolute time is a geodesic of ∇ if and only if

$$(\overset{v}{\nabla}_{\dot\gamma}\dot\gamma)^v = -\overset{v}{\nabla}{}^v\phi \tag{3.71}$$

along γ, and

(ii) any such ϕ satisfies Poisson's equation

$$\overset{v}{\nabla}_\mu \overset{v}{\nabla}{}^\mu \phi = 4\pi G\rho, \tag{3.72}$$

where ρ is the mass density.

Proof We apply the Künzle–Ehlers recovery theorem (Theorem 3.28) in the case $\omega = 0$. The recovered field equations (3.49) reduce to

$$\mathrm{d}(\alpha|_\Sigma) = 0, \tag{3.73a}$$
$$-\,{}^{(n)}\delta(\alpha|_\Sigma) = 4\pi G\rho. \tag{3.73b}$$

The first of these tells us that on each spatial leaf Σ, $\alpha|_\Sigma$ is closed, such that by the Poincaré lemma we locally have $\alpha|_\Sigma = \mathrm{d}(\phi_\Sigma)$ for some function ϕ_Σ on Σ. Combining those for different Σ, we obtain a (locally defined) function ϕ on M which satisfies[13]

$$\alpha|_\Sigma = \mathrm{d}(\phi|_\Sigma). \tag{3.74}$$

Inserting this into (3.73b), we obtain Poisson's equation; and inserting it into the test particle equation of motion (3.48) from the Künzle–Ehlers recovery theorem, we obtain the equation of motion (3.71).

Furthermore, *any* function ϕ for which the equation of motion takes the form (3.71) must satisfy (3.74). This determines ϕ uniquely up to addition of a spatially constant function. $\square$

---

[13] Note that (3.74) means that as a form on spacetime, we have $\alpha = \mathrm{d}\phi + f\tau$ for some function f, which by $\alpha_\mu v^\mu = 0$ is fixed to be $f = -v(\phi)$.

Remark 3.41 In coordinates adapted to v as in Construction 3.29, this is the standard formulation of Newtonian gravity with gravitational potential ϕ: The test particle equation of motion becomes

$$\ddot{\gamma}^a = -\delta^{ab}\partial_b\phi, \tag{3.75}$$

and the field equation for the potential ϕ becomes

$$\delta^{ab}\partial_a\partial_b\phi = 4\pi G\rho. \tag{3.76}$$

$$\triangle$$

Exercise 3.42 (*Newtonian potential under change of frame*)   In the Trautman recovery theorem, the Newtonian potential ϕ with respect to a twist-free rigid unit timelike vector field v is defined (up to addition of spatially constant functions) by the acceleration α of v satisfying

$$\alpha|_\Sigma = \mathrm{d}(\phi|_\Sigma), \tag{3.77a}$$

textor in components

$$\alpha^\mu = \nabla^\mu\phi. \tag{3.77b}$$

Let $\tilde{v}$ be another twist-free rigid field, and let w be the Milne boost field relating the two fields v and $\tilde{v}$, i.e. $w := v - \tilde{v}$. Show that a function $\tilde{\phi}$ is a Newtonian potential for $\tilde{v}$ if and only if it is related to ϕ by

$$\nabla^\mu(\phi - \tilde{\phi}) = (\overset{v}{\nabla}_v w)^\mu. \tag{3.78}$$

What does this equation look like in coordinates adapted to v?

Hint: Use the results of Exercise 3.30 *on the behaviour of acceleration and twist of rigid fields under Milne boosts.* $\triangle$

As was shown by Ehlers (fleshing out an idea by Künzle), the absolute rotation condition follows from a weak notion of asymptotic flatness:

Proposition 3.43 *Let (M, τ, h, ∇) be a Newtonian manifold that satisfies the Newton–Cartan field equation and is spatially flat, and whose spatial leaves are simply connected and geodesically complete. If for any spacelike geodesic $\gamma : \mathbb{R} \to M$, we have*

$$\left(R^\mu{}_{\nu\kappa\rho}R^\nu{}_\mu{}^\kappa{}_\sigma v^\rho v^\sigma\right)\Big|_{\gamma(s)} \xrightarrow{s\to\infty} 0 \tag{3.79}$$

for some unit timelike vector field v, then the manifold has absolute rotation.

Proof Each spatial leaf is a simply connected geodesically complete flat Riemannian manifold of dimension n, i.e. isometric to n-dimensional Euclidean space.

Due to spatial flatness, the expression on the left-hand side of (3.79) is the same for *any* unit timelike vector field v. In particular, for v rigid, by Corollary 3.16 we obtain

$$\left.\left((\nabla_\kappa \omega^\mu{}_\nu)\nabla^\kappa \omega^\nu{}_\mu\right)\right|_{\gamma(s)} \xrightarrow{s\to\infty} 0, \tag{3.80}$$

i.e. $\overset{(n)}{\nabla}\omega$ goes to zero at infinity on each spatial leaf.

Now by the Künzle–Ehlers recovery theorem, on each spatial leaf Σ, ω satisfies $d(\omega|_\Sigma) = 0$ and $^{(n)}\delta(\omega|_\Sigma) = 0$, i.e. $\omega|_\Sigma$ is a harmonic form. This implies that its components ω_{ab} with respect to orthonormal coordinates are harmonic functions, which in turn implies that $\partial_a \omega_{bc}$ are also harmonic functions. Since those vanish at infinity, by the maximum principle for harmonic functions on $\mathbb{R}^n$ they vanish on all of Σ. Thus ω is spatially constant, and we are done. $\qquad\square$

Exercise 3.44 (*Newtonian cosmology—inconsistent?*)[14]　　We consider standard Newtonian gravity (in n spatial dimensions) in its usual coordinate formulation; i.e. we have a gravitational potential ϕ satisfying Poisson's equation

$$\Delta\phi = 4\pi G\rho, \tag{3.81}$$

where $\Delta = \sum_{a=1}^{n} \partial_a^2$ is the Laplacian on n-dimensional Euclidean space and ρ is the mass density. We want to analyse Newtonian *cosmology* for a homogeneous isotropic universe, i.e. we assume ρ to be spatially constant.

(a) Show that the Newtonian gravitational field $g_a = -\partial_a \phi$ in general *cannot* be homogeneous.

　　Hint: If it were homogeneous, what would Poisson's equation imply?

(b) Show that, given any point $\vec{p} \in \mathbb{R}^n$, any function ϕ satisfying

$$\partial_a \phi(\vec{x}) = \frac{4}{n}\pi G\rho(x_a - p_a) \tag{3.82}$$

　　solves Poisson's equation. Such a solution we call a *canonical solution centred at $\vec{p}$*.

(c) By explicitly writing down a canonical solution centred at $\vec{p}$, show that canonical solutions exist.

　　Hint: You should be able to guess a solution.

(d) Show that, if $\rho > 0$, ϕ being a canonical solution centred at some point (i.e. satisfying (3.82) for some $\vec{p}$) is equivalent to

$$\partial_a \partial_b \phi = \frac{4}{n}\pi G\rho\,\delta_{ab}. \tag{3.83}$$

　　Hint: Consider the difference of ϕ to a canonical solution $\tilde{\phi}$ centred at some $\tilde{\tilde{p}}$.

---

[14] This and the following two exercises are based on work by Malament, see ref. [7, Sect. 4.4].

That the gravitational field in a homogeneous Newtonian universe cannot be homogeneous was in the past considered as evidence that the notion of Newtonian cosmology is inconsistent. The existence of several solutions centred at *any* point, towards which the gravitational field points, seems to make this even worse—which point should be the preferred one?

As it turns out, this apparent problem can be solved by realising that the Newtonian gravitational field $-\partial_a \phi$ can never be measured on its own: one always measures the combination of gravitational and inertial forces (in the standard Newtonian sense), and one can show that this leads to canonical solutions centred at any two different points being empirically indistinguishable.

In Newton–Cartan gravity, as in GR gravity is not considered a force, but unified with inertia. Therefore, one might expect the above problem to have a natural solution in the context of Newton–Cartan gravity. In the following two exercises, we will see that this is indeed the case. $\triangle$

Exercise 3.45 (*FLRW-like Newton–Cartan cosmology*) By a *Newton–Cartan cosmological model*, we will mean a Galilei manifold (M, τ, h) with absolute time with a torsion-free Galilei connection ∇, together with a unit timelike vector field ξ, interpreted as the spacetime velocity[15] of the cosmological matter, and a function ρ, interpreted as the mass density of the cosmological matter, such that the Newton–Cartan field equation $R_{\mu\nu} = 4\pi G \rho \tau_\mu \tau_\nu$ is satisfied. Note that due to the field equation ρ is redundant, since it is determined by the other data of the model.

A *symmetry* of a cosmological model $(M, \tau, h, \nabla, \xi)$ is a diffeomorphism $\varphi \colon M \to M$ that leaves all of the data invariant, i.e. satisfies

$$\varphi^* \tau = \tau, \quad \varphi^* h = h, \quad \varphi^* \nabla = \nabla, \quad \varphi^* \xi = \xi. \tag{3.84}$$

The model is *(spatially) homogeneous and isotropic* if for any point $p \in M$ and any spatial rotation[16] $R \in \mathsf{SO}(\ker \tau|_p, {}^{(n)}h)$ there exists a symmetry φ with $\varphi(p) = p$ and $(D\varphi|_p)|_{\ker \tau|_p} = R$ ('all spatial rotations are realisable by symmetries').[17]

Let $(M, \tau, h, \nabla, \xi)$ be a homogeneous and isotropic Newton–Cartan cosmological model.

---

[15] Since we don't restrict to the case of four spacetime dimensions, we won't call this the 'four-velocity'.

[16] Even though we have no canonical orientation on the vector space $\ker \tau|_p$ of spacelike vectors at p, the notion of a linear automorphism of $\ker \tau|_p$ being orientation-preserving or -reversing is still well-defined. Thus it makes sense to speak of the group $\mathsf{SO}(\ker \tau|_p, {}^{(n)}h)$ of (proper) spatial rotations at p, and not only of the group $\mathsf{O}(\ker \tau|_p, {}^{(n)}h)$ of improper spatial rotations.

[17] In $n = 3$ spatial dimensions, the condition that each spatial rotation be realisable by a symmetry is equivalent to demanding that for any two unit spacelike vectors at a point there exist a symmetry mapping one onto the other ('no preferred spacelike directions'). However, in higher dimensions this equivalence fails: for $n > 4$, there are proper subgroups of $\mathsf{SO}(n)$ whose action on S^{n-1} is transitive.

(a) Show that any vector field determined by τ, h, ∇, ξ (in terms of tensor products, contraction, and covariant and exterior differentiation) is proportional to ξ.
Hint: Such a vector field is invariant under all symmetries of the model. It can be written as $a\xi + w$ with w spacelike, and then isotropy implies …

(b) Deduce that ξ is geodesic, and that ρ is spatially constant.

(c) Show that any spacelike symmetric contravariant degree-2 tensor field $\lambda = \lambda^{\mu\nu}\partial_\mu\partial_\nu$ determined by τ, h, ∇, ξ (in the above sense) is proportional to h.
Hint: At any point p, λ can be expressed as $\lambda|_p = \sum_{a=1}^{n} b^a \mathrm{e}_a \otimes \mathrm{e}_a$, where the e_a form an orthonormal basis of spacelike vectors. To this, apply isotropy.

(d) Deduce that ξ has vanishing shear.

(e) Show that for $n > 2$, any purely spacelike 2-form λ determined by τ, h, ∇, ξ (in the above sense) vanishes.
Hint: At any point p, λ can be expressed as $\lambda|_p = \sum_{a<b}^{n} \lambda_{ab}\mathrm{e}^a \wedge \mathrm{e}^b$, where the e^a are dual to an orthonormal basis of spacelike vectors. Consider isotropy for rotations that swap two basis vectors e_c and e_d.

(f) Deduce that ξ has vanishing twist.

(g) Why do the preceding statements show that ∇ is Newtonian?
Hint: Newton–Coriolis form!

(h) Assuming that the spacetime is spatially flat, show that it has absolute rotation.
Hint: Construct a rigid twist-free vector field v from ξ, by 'subtracting the expansion'. This works similar to the construction of twist-free vector fields in Proposition 3.39.

We thus have shown that in any homogeneous and isotropic Newton–Cartan cosmological model, the spacetime velocity of the cosmological matter is geodesic, shear- and twist-free, and its mass density is spatially constant. Furthermore, we showed that spacetime is Newtonian (i.e. that we have a proper model of Newton–Cartan gravity), and that assuming spatial flatness we get absolute rotation. $\triangle$

Exercise 3.46 (*Newtonian cosmology—not inconsistent!*) Let $(M, \tau, h, \nabla, \xi)$ be a spatially flat homogeneous and isotropic cosmological Newton–Cartan model as defined in Exercise 3.45, and let ρ be the corresponding mass density function. Here we are going to show that the solutions of ordinary Newtonian gravity we get from this by applying the Trautman recovery theorem are precisely the canonical solutions to Newtonian cosmology we know from Exercise 3.44.

(a) Let v be *any* rigid twist-free vector field, and let ϕ be a corresponding Newtonian gravitational potential from the Trautman recovery theorem. Show that it satisfies

$$\nabla^\mu\nabla^\nu\phi = \frac{4}{n}\pi G\rho h^{\mu\nu}, \tag{3.85}$$

where $n = \dim M - 1$.
Hint: Express the fact that ξ is geodesic (which we know from Exercise 3.45 (b)) via the Trautman recovery theorem, and use this to compute $\nabla^\mu\nabla^\nu\phi = \overset{v}{\nabla}{}^\mu\overset{v}{\nabla}{}^\nu\phi$.

Then express $\overset{v}{\nabla}_\xi \xi$ in terms of $\nabla_\xi \xi$, rewrite this in terms of the expansion of ξ, and use the Raychaudhuri equation (3.12).

(b) Using the frame change result of Exercise 3.42, show that for *any* function $\tilde{\phi}$ satisfying (3.85), there is a rigid twist-free vector field $\tilde{v}$ such that $\tilde{\phi}$ is a Newtonian potential with respect to $\tilde{v}$ in the sense of the Trautman recovery theorem.
Hint: Consider the difference $\phi - \tilde{\phi}$ to construct the Milne boost relating v to $\tilde{v}$.

Equation (3.85) is a purely spacelike equation, independent on each spatial leaf. In orthonormal coordinates on any spatial leaf, it takes the form (3.83), which we have shown in Exercise 3.44 (d) to be (for $\rho > 0$) equivalent to ϕ being a canonical solution to Poisson's equation (on the corresponding spatial leaf). So we have in fact recovered the canonical solutions of Newtonian cosmology, which in the standard formulation are seemingly non-homogeneous and non-isotropic, from a perfectly homogeneous and isotropic cosmological model. $\triangle$

3.4 Newton–Cartan Gravity as the Newtonian Limit of General Relativity

In this section, we will show in which sense Newton–Cartan gravity arises as the 'Newtonian limit' of GR.[18] We will formulate the Newtonian limit as the formal limit $c \to \infty$, where c is the speed of light. Concretely, we will implement this limit by expanding the objects appearing in GR as power series in c^{-1}, which we thereby treat as a 'small parameter'. Intuitively, when taking the limit $c \to \infty$, the objects with which we probe a general-relativistic spacetime become slower and slower as compared to the speed of light. Of course, to make sense of this, we need a set of reference observers throughout spacetime with respect to which we define 'slow objects': *a Newtonian limit only makes sense with respect to a reference of 'slow motion'*. Pictorially, as we take $c \to \infty$, the lightcones of the original Lorentzian manifold 'open up' towards the rest spaces of these reference observers, and we end up with a Galilei manifold with its distribution of spacelike vectors; see Fig. 3.4.

Of course, analytically speaking, a 'Taylor expansion' in a dimensionful parameter like c does not make sense (even more so since c is a constant of nature); only for *dimensionless* parameters can a meaningful 'small-parameter approximation' be made. In physical realisations of the limit from GR to Newton–Cartan gravity, this means that the corresponding small parameter has to be chosen as, e.g., the ratio of some typical velocity of the system under consideration to the speed of light. Ehlers discusses the Newtonian limit from GR to Newton–Cartan gravity in terms of an actual small parameter approaching zero [2, 3], and other places in the literature

---

[18] Some of the following preliminary discussion is taken from my (the author's) article [9].

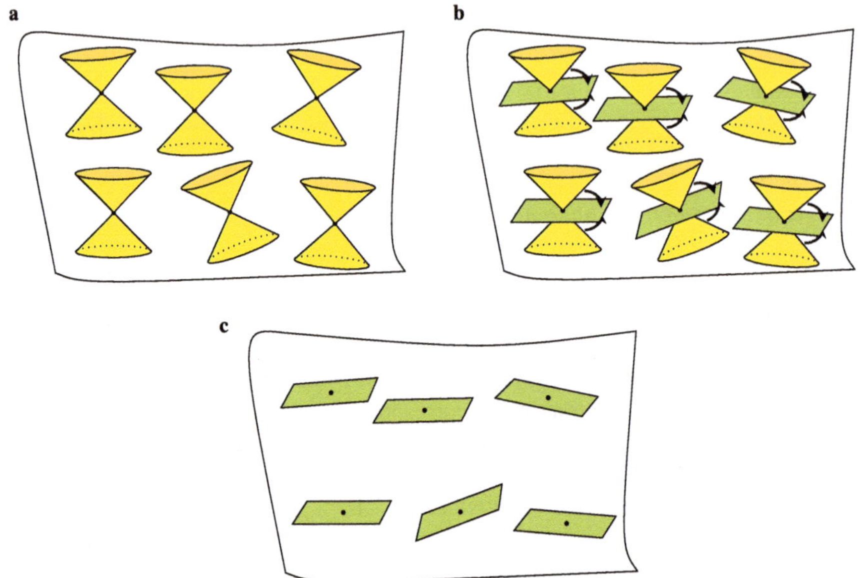

Fig. 3.4 The Newtonian limit. **a** Step 1: a Lorentzian manifold. **b** Step 2: The lightcones opening up as $c \to \infty$, towards the rest spaces with respect to a reference state of motion. **c** Step 3: the resulting Galilei manifold

discuss the relationship of formal '$c \to \infty$' limits to actual physical approximations, see, e.g., ref. [12, Sect. II].

In the following, however, we will forget about these issues and treat the 'small c^{-1} expansion' in a purely formal way: all expansions in c^{-1} will be treated as formal power series—or, more precisely, formal Laurent series, since we will need negative orders of c^{-1} as well.[19] As we will see, this expansion exhibits Lorentzian geometry as a formal deformation of Galilei geometry, and GR as a formal deformation of Newton–Cartan gravity. For a quantity X being of order (at least) k in the formal c^{-1}-expansion, we will use the notation

$$X = \mathrm{O}(c^{-k}), \tag{3.86}$$

---

[19] Strictly speaking, this means that we are not dealing with tensor fields (or more general differential-geometric objects) in the usual sense, but with tensor fields (or more general geometric objects) that take values in the field of formal Laurent series $\mathbb{R}((c^{-1}))$ instead of the real numbers. However, for the formal treatment of the theory this does not cause any problems—we just define differentiation of series-valued objects order by order, and demand that equations be satisfied order by order. We will also need to compare elements of $\mathbb{R}((c^{-1}))$ in size. For this, we equip $\mathbb{R}((c^{-1}))$ with the natural order, namely the unique total order making it into an ordered field that on $\mathbb{R} \subset \mathbb{R}((c^{-1}))$ agrees with the standard order and satisfies $0 < c^{-1} < \varepsilon$ for all $\varepsilon \in \mathbb{R}$, $\varepsilon > 0$. Explicitly, this means that an element of $\mathbb{R}((c^{-1}))$ is considered positive if (and only if) its leading-order term in the c^{-1}-expansion is positive.

which—to stress it again—is not to be understood as any analytic statement. If some quantity has no non-vanishing terms of negative order in c^{-1}, i.e. its expansion is a proper formal power series (and not only a Laurent series), then it has a 'formal $c \to \infty$ limit', namely the term of order c^0.

Construction 3.47 When taking the Newtonian limit $c \to \infty$, the rest spaces of our reference observers, with respect to which we define the notion of 'slowness', become the spaces $\ker \tau|_p$ of spacelike vectors in the resulting Galilei manifold. Therefore, the choice of reference observers is contained in the choice of the one-form τ, with physical dimension of time, whose kernels are the observers' rest spaces.[20]

The Lorentzian temporal length of a timelike vector v with respect to the metric g is $c^{-1}\sqrt{-g(v, v)}$, and in the Newtonian limit $c \to \infty$ we want this to become the temporal length $\tau(v)$ in the Newton–Cartan sense: we want $c^{-1}\sqrt{-g(v, v)} = \tau(v) + \mathrm{O}(c^{-1})$. From this, it follows that the metric has to satisfy $g = -c^2\tau \otimes \tau + \mathrm{O}(c^1)$.

In order to obtain a satisfactory Newtonian limit, we will also assume that the inverse metric has no terms of negative order in c^{-1}, i.e. that $g^{-1} = \mathrm{O}(c^0)$. Further, we make the additional simplifying assumption that the next-to-leading-order term in the expansion of the metric is of order at least c^0, i.e. the c^1 term vanishes and we have $g = -c^2\tau \otimes \tau + \mathrm{O}(c^0)$.[21] Due to this assumption, the term of order c^{-1} in the expansion of the inverse metric g^{-1} will drop out of all of the following considerations; so without loss of generality we will assume it to vanish.

Put differently, in our consideration of Newtonian limits we will assume existence of the formal '$c \to \infty$ limits'

$$\lim_{c\to\infty} (c^{-2}g) = -\tau \otimes \tau, \quad \lim_{c\to\infty} g^{-1}, \tag{3.87}$$

as well as vanishing of the order-c^1 term of g and the order-c^{-1} term of g^{-1}.[22]   △

Lemma 3.48 *Let (M, g) be a Lorentzian manifold, and assume that the metric g may be expanded as a formal power series[23] in c^{-1} as*

$$g = -c^2\tau \otimes \tau + \mathrm{O}(c^0) \tag{3.88a}$$

---

[20] The Lorentzian spacetime velocity of the observers will then be given by the vector field $V = -cg^{-1}(\tau, \cdot)/\sqrt{-g^{-1}(\tau, \tau)}$, where the prefactor c is for dimensional reasons, and the minus sign is for the time orientation of τ to agree with that of V.

[21] A discussion of (pseudo-)Newtonian limits where this assumption is dropped may be found in the article [4].

[22] For a similar but slightly different set of assumptions for the Newtonian limit, starting from the inverse metric instead of the metric, see ref. [6, Sect. 5] or ref. [12, Sect. II A]. A combined discussion of these different possible assumptions for the Newtonian limit may also be found in my (the author's) article [10, Proposition 3.1]. The proof of Lemma 3.48 (i) is in parallel to that given in this article.

[23] As said above, this is a formal *Laurent* series, since we have a term of negative order. We will however continue to use the term 'power series', since most of our series will have terms of only non-negative order.

for some nowhere-vanishing one-form $\tau \in \Omega^1(M)$. Assume further that the inverse metric has the expansion

$$g^{-1} = h + O(c^{-2}) \tag{3.88b}$$

for some contravariant degree-2 tensor field h.

(i) *(M, τ, h) is a Galilei manifold.*
(ii) *Writing $g^{-1} =: h + c^{-2}m + O(c^{-3})$, the vector field $\hat{v} := -m(\tau, \cdot)$ is unit time-like in the Newton–Cartan sense (with respect to τ).*
(iii) *Using $\hat{v}$ from above, the metric has the expansion*

$$g = -c^2\tau \otimes \tau + \underset{\hat{v}}{h} - 2\hat{\phi}\tau \otimes \tau + O(c^{-1}) \tag{3.89}$$

for some function $\hat{\phi}$.

Proof We write the expansion of the metric as

$$g = -c^2\tau \otimes \tau + g^{(0)} + O(c^{-1}) \tag{3.90}$$

for some as of yet unknown tensor field $g^{(0)}$.

The definition of the inverse metric reads $g_{\mu\nu}g^{\nu\rho} = \delta^\rho_\mu$, which with the assumed expansions becomes

$$\delta^\rho_\mu = -c^2\tau_\mu\tau_\nu h^{\nu\rho} + g^{(0)}_{\mu\nu}h^{\nu\rho} - \tau_\mu\tau_\nu m^{\nu\rho} + O(c^{-1}). \tag{3.91}$$

Comparison of coefficients implies $\tau_\nu h^{\nu\rho} = 0$ and

$$\delta^\rho_\mu = g^{(0)}_{\mu\nu}h^{\nu\rho} - \tau_\mu\tau_\nu m^{\nu\rho}. \tag{3.92}$$

Since the identity map has rank $\dim M =: n + 1$ and the last term in (3.92) has rank 1,[24] the first term on the right-hand side has rank $\geq n$. This implies that h has rank $\geq n$; since it is degenerate in direction τ, it has rank n. To show that (M, τ, h) is a Galilei manifold, it remains to prove that in its n non-degenerate directions h is positive definite. This we will do in a moment.

Contracting (3.92) with τ_ρ, we obtain $-1 = \tau_\nu m^{\nu\rho}\tau_\rho$. This shows on the one hand that $\hat{v} := -m(\tau, \cdot)$ is unit timelike (in the Newton–Cartan sense), and on the other hand that $g^{-1}(\tau, \tau) = -c^{-2} + O(c^{-3}) < 0$, i.e. that τ is timelike in the Lorentzian sense.

Let now β be any covector such that $h(\beta, \beta) \neq 0$. This implies that β and τ are linearly independent, and therefore the projection $\tilde{\beta} := \beta - \frac{g^{-1}(\tau,\beta)}{g^{-1}(\tau,\tau)}\tau$ of β onto the

---

[24] If it had rank 0, the first term on the right-hand side would be of rank $n + 1$, which can't be the case since h has a degenerate direction and therefore rank at most n.

g^{-1}-orthogonal complement of τ is nonzero. Since τ is timelike, $\tilde{\beta}$ is spacelike (both in the Lorentzian sense). One easily computes that $\tilde{\beta} = \beta + m(\tau, \beta)\tau + O(c^{-1})$, which implies $0 < g^{-1}(\tilde{\beta}, \tilde{\beta}) = h(\beta, \beta) + O(c^{-1})$. Thus $h(\beta, \beta) > 0$, and we have shown that h is positive definite in its non-degenerate directions. Thus now we know that (M, τ, h) is a Galilei manifold.

Now comparing (3.92) to the definition $\delta_\mu^\rho = P_\mu^\rho + \tau_\mu \hat{v}^\rho$ of the spatial projector along $\hat{v}$, we obtain $g_{\mu\nu}^{(0)} h^{\nu\rho} = P_\mu^\rho$. This in turn implies $g_{\mu\nu}^{(0)} P_\rho^\nu = \hat{h}_{\mu\rho}$, where $\hat{h}_{\mu\rho}$ are the components of $\underset{\hat{v}}{h}$. Thus we finally obtain

$$g_{\mu\nu}^{(0)} = g_{\rho\sigma}^{(0)}(P_\mu^\rho + \hat{v}^\rho \tau_\mu)(P_\nu^\sigma + \hat{v}^\sigma \tau_\nu) = \hat{h}_{\mu\nu} + g^{(0)}(\hat{v}, \hat{v})\tau_\mu \tau_\nu \qquad (3.93)$$

and are finished with $\hat{\phi} = -\tfrac{1}{2}g^{(0)}(\hat{v}, \hat{v})$. $\qquad\qquad\square$

Construction 3.49 We will now compute the Christoffel symbols of the Lorentzian metric g, i.e. the connection coefficients of the Levi-Civita connection. Using the expansions from Lemma 3.48, the Christoffel symbols are

$$
\begin{aligned}
\overset{g}{\Gamma}{}^\rho_{\mu\nu} &= \frac{1}{2}g^{\rho\sigma}(2\partial_{(\mu}g_{\nu)\sigma} - \partial_\sigma g_{\mu\nu}) \\
&= \frac{1}{2}\big(h^{\rho\sigma} + c^{-2}m^{\rho\sigma} + O(c^{-3})\big) \\
&\quad \big(-(c^2 + 2\hat{\phi})2(\tau_\sigma \partial_{(\mu}\tau_{\nu)} + \tau_{(\nu}\partial_{\mu)}\tau_\sigma - \tau_{(\mu}\partial_{|\sigma|}\tau_{\nu)}) \\
&\quad + 2\partial_{(\mu}\hat{h}_{\nu)\sigma} - \partial_\sigma \hat{h}_{\mu\nu} - 4\tau_\sigma \tau_{(\nu}\partial_{\mu)}\hat{\phi} + 2\tau_\mu \tau_\nu \partial_\sigma \hat{\phi} + O(c^{-1})\big) \\
&= -(c^2 + 2\hat{\phi})h^{\rho\sigma}(\tau_{(\nu}\partial_{\mu)}\tau_\sigma - \tau_{(\mu}\partial_{|\sigma|}\tau_{\nu)}) + \frac{1}{2}h^{\rho\sigma}(2\partial_{(\mu}\hat{h}_{\nu)\sigma} - \partial_\sigma \hat{h}_{\mu\nu}) \\
&\quad + h^{\rho\sigma}\tau_\mu \tau_\nu \partial_\sigma \hat{\phi} - m^{\rho\sigma}(\tau_{(\nu}\partial_{\mu)}\tau_\sigma - \tau_{(\mu}\partial_{|\sigma|}\tau_{\nu)}) - m^{\rho\sigma}\tau_\sigma \partial_{(\mu}\tau_{\nu)} + O(c^{-1}) \\
&= -[(c^2 + 2\hat{\phi})h^{\rho\sigma} + m^{\rho\upsilon}]\tau_{(\mu}(d\tau)_{\nu)\sigma} + \hat{v}^\rho \partial_{(\mu}\tau_{\nu)} \\
&\quad + \frac{1}{2}h^{\rho\sigma}(2\partial_{(\mu}\hat{h}_{\nu)\sigma} - \partial_\sigma \hat{h}_{\mu\nu}) + h^{\rho\sigma}\tau_\mu \tau_\nu \partial_\sigma \hat{\phi} + O(c^{-1}). \qquad (3.94)
\end{aligned}
$$

$\triangle$

Theorem 3.50 *Let (M, g) be a Lorentzian manifold whose metric and inverse metric have expansions as in Lemma 3.48.*

(i) (M, τ, h) is a Galilei manifold.

(ii) Any worldline γ in M that is timelike in the Lorentzian sense is also timelike in the Galilei sense. If $\tau(\dot{\gamma}) > 0$ along γ, in the Newtonian limit the Lorentzian proper time along γ goes over to time in the Galilei sense.

(iii) Let w be a 'potentially c-dependent vector' with a regular formal $c \to \infty$ limit, i.e. $w \in TM \otimes \mathbb{R}[[c^{-1}]]$. If w is spacelike in the Lorentzian sense, its limit $\tilde{w} = \lim_{c \to \infty} w$ is spacelike in the Galilei sense, and the Lorentzian metric length $\sqrt{g(w, w)}$ of w in the limit goes over to the Galilei metric length $\sqrt{^{(n)}h(\tilde{w}, \tilde{w})}$

of $\tilde{w}$. Put differently, spatial lenghts as defined by g in the limit go over to spatial lengths as defined by h.

(iv) The Levi-Civita connection $\overset{g}{\nabla}$ of (M, g) has a regular formal $c \to \infty$ limit, i.e. no terms of negative order in c^{-1}, if and only if $d\tau = 0$.

(v) If $d\tau = 0$, the formal $c \to \infty$ limit of $\overset{g}{\nabla}$, i.e. its term of order c^0, is a torsion-free Galilei connection ∇ on (M, τ, h). The curvature tensors of $\overset{g}{\nabla}$ and of ∇ satisfy the formal limit relation

$$\lim_{c \to \infty} \overset{g}{R}{}^{\mu}{}_{\nu\rho\sigma} = R^{\mu}{}_{\nu\rho\sigma} \, , \tag{3.95}$$

and ∇ is Newtonian.

Proof (i) This was proved in Lemma 3.48.

(ii) From Lemma 3.48, we know that $g = -c^2\tau \otimes \tau + \underset{\hat{v}}{h} - 2\hat{\phi}\tau \otimes \tau + \mathrm{O}(c^{-1})$. Therefore, if a vector $w \in TM$, $w \neq 0$, is spacelike in the Galilei sense, we have $g(w, w) = \underset{\hat{v}}{h}(w, w) + \mathrm{O}(c^{-1}) > 0$, such that it is also Lorentzian spacelike. Conversely, any $w \in TM$, $w \neq 0$ that is causal in the Lorentzian sense must be timelike in the Galilei sense.[25]

For a timelike curve γ with $\tau(\dot{\gamma}) > 0$, the proper time expands as

$$c^{-1} \int_{\gamma} \sqrt{-g_{\mu\nu}\mathrm{d}x^{\mu}\mathrm{d}x^{\nu}} = c^{-1} \int_{\gamma} \sqrt{c^2\tau_{\mu}\tau_{\nu}\mathrm{d}x^{\mu}\mathrm{d}x^{\nu} + \mathrm{O}(c^0)} = \int_{\gamma} \tau + \mathrm{O}(c^{-2}). \tag{3.96}$$

(iii) We have $w = \tilde{w} + \mathrm{O}(c^{-1})$. Again due to $g_{\mu\nu} = -c^2\tau_{\mu}\tau_{\nu} + \hat{h}_{\mu\nu} - 2\hat{\phi}\tau_{\mu}\tau_{\nu} + \mathrm{O}(c^{-1})$, if $\tilde{w}$ were timelike in the Galilei sense, i.e. $\tau(\tilde{w}) \neq 0$, we would have $g(w, w) = -c^2(\tau(\tilde{w}))^2 + \mathrm{O}(c^1) < 0$, i.e. w would be Lorentzian timelike. Therefore for w Lorentzian spacelike, we have $\tau(\tilde{w}) = 0$. For the length, we then directly compute

$$\sqrt{g(w, w)} = \sqrt{\underset{\hat{v}}{h}(\tilde{w}, \tilde{w}) + \mathrm{O}(c^{-1})} = \sqrt{\underset{\hat{v}}{h}(\tilde{w}, \tilde{w})} + \mathrm{O}(c^{-1})$$

$$= \sqrt{{}^{(n)}h(\tilde{w}, \tilde{w})} + \mathrm{O}(c^{-1}). \tag{3.97}$$

(iv) By decomposing into spacelike and timelike components, one easily checks that $\tau_{(\mu}(\mathrm{d}\tau)_{\nu)}{}^{\sigma} = 0$ is equivalent to $\mathrm{d}\tau = 0$. Hence, according to the calculation in Construction 3.49, the order-c^2 term of the Christoffel symbols of g—which is the only negative-order term—vanishes if and only if $\mathrm{d}\tau = 0$.

---

[25] Since any $w \in TM$ is c-independent, it cannot be lightlike.

(v) Comparing the limiting connection from Construction 3.49 in the case $d\tau = 0$ to the general form of Galilei connections from the classification theorem (Theorem 2.29), we see that the limit is a torsion-free Galilei connection.

Since the connection coefficients satisfy $\overset{g}{\Gamma}{}^{\rho}_{\mu\nu} = \Gamma^{\rho}_{\mu\nu} + O(c^{-1})$ and curvature tensor components have the symbolic form $R = \partial\Gamma - \partial\Gamma + \Gamma\Gamma - \Gamma\Gamma$, we directly obtain (3.95). Finally, from

$$\overset{g}{R}{}^{\mu}{}_{\rho}{}^{\nu}{}_{\sigma} = g^{\nu\kappa}\overset{g}{R}{}^{\mu}{}_{\rho\kappa\sigma} = h^{\nu\kappa}R^{\mu}{}_{\rho\kappa\sigma} + O(c^{-1}) = R^{\mu}{}_{\rho}{}^{\nu}{}_{\sigma} + O(c^{-1}) \qquad (3.98)$$

and symmetry in pairs of the Riemannian curvature tensor, we obtain symmetry in pairs for $R^{\mu}{}_{\rho}{}^{\nu}{}_{\sigma}$, i.e. that ∇ is Newtonian. $\qquad\square$

Remark 3.51 Comparing the connection coefficients from Construction 3.49 to the classification theorem, we obtain even more: the Newton–Coriolis form $\hat{\Omega}$ of the limiting Galilei connection ∇ with respect to the unit timelike vector field $\hat{v} = -m(\tau, \cdot)$ satisfies

$$\tau_{(\mu}\hat{\Omega}_{\nu)}{}^{\rho} = h^{\rho\sigma}\tau_{\mu}\tau_{\nu}\partial_{\sigma}\hat{\phi}. \qquad (3.99a)$$

By decomposing into spacelike and timelike parts, from this we obtain

$$\hat{\Omega}_{\mu\nu} = 2\tau_{[\mu}\partial_{\nu]}\hat{\phi} \quad \text{or, in component-free notation,} \quad \hat{\Omega} = \tau \wedge d\hat{\phi}. \qquad (3.99b)$$

Therefore, the unit timelike vector field $\hat{v}$ obtained from the expansion of the Lorentzian metric is twist-free and has acceleration $\hat{\alpha} = d\hat{\phi}$. However, $\hat{v}$ is not necessarily rigid. $\qquad\triangle$

Construction 3.52 We now want to look at the Newtonian limit of the Einstein equation

$$\overset{g}{R}_{\mu\nu} - \frac{1}{2}\overset{g}{R}g_{\mu\nu} = \frac{n-1}{n-2}\frac{4\pi G}{c^4}T_{\mu\nu}. \qquad (3.100a)$$

Here the prefactor, where $n = \dim M - 1 \neq 2$, is purely conventional; in $n = 3$ spatial dimensions it gives the usual $8\pi G/c^4$. We have chosen this prefactor such that in the Newtonian limit we will obtain the prefactor in the Newton–Cartan field equations as introduced before.

In order to be able to obtain a regular Newtonian limit, we have to consider the Einstein equation's trace-reversed form[26]

$$\overset{g}{R}_{\mu\nu} = \frac{n-1}{n-2}\frac{4\pi G}{c^4}\left(T_{\mu\nu} - \frac{1}{n-1}Tg_{\mu\nu}\right), \qquad (3.100b)$$

---

[26] The name 'trace-reversed' is conventional. In $n \neq 3$ spatial dimensions, this form does not actually have a reversed trace compared to the original one, in the sense of a sign flip; instead, the trace is multiplied by a factor of $1 - \frac{n+1}{n-1} = -\frac{2}{n-1}$.

where we now additionally have to assume $n \neq 1$, and $T = g^{\mu\nu}T_{\mu\nu}$ is the trace of the energy–momentum tensor.

We assume the metric and inverse metric to have expansions as in Lemma 3.48. Defining the Lorentzian normalised one-form $V_\mu := \tau_\mu/\sqrt{-g^{-1}(\tau,\tau)}$, we assume the energy–momentum tensor to have the expansion

$$T_{\mu\nu} = w V_\mu V_\nu + 2c\Pi_{(\mu} V_{\nu)} + \Sigma_{\mu\nu} \tag{3.101a}$$

where w, Π, Σ are, respectively, the energy density, the momentum density (energy flux density), and the momentum flux density (stress tensor) with respect to an observer with spacetime velocity $-cg^{-1}(V,\cdot)$. We expand the energy density as

$$w = \rho c^2 + \mathrm{O}(c^1) \tag{3.101b}$$

in terms of the (rest) mass density ρ, and we assume momentum density and momentum flux density to be of order c^0. The Lorentzian length of τ is $\sqrt{-g^{-1}(\tau,\tau)} = \sqrt{-h(\tau,\tau) - c^{-2}m(\tau,\tau) + \mathrm{O}(c^{-3})} = c^{-1}\sqrt{1+\mathrm{O}(c^{-1})} = c^{-1}(1+\mathrm{O}(c^{-1}))$. This implies $V_\mu = c\tau_\mu(1+\mathrm{O}(c^{-1}))^{-1} = c\tau_\mu(1+\mathrm{O}(c^{-1}))$. Using these expansions, we obtain

$$T_{\mu\nu} = (\rho c^4 + \mathrm{O}(c^3))\tau_\mu\tau_\nu + (c^2 + \mathrm{O}(c^1))2\Pi_{(\mu}\tau_{\nu)} + \mathrm{O}(c^0). \tag{3.101c}$$

Contracting this with $g^{\mu\nu} = h^{\mu\nu} + c^{-2}m^{\mu\nu} + \mathrm{O}(c^{-3})$, the trace of the energy–momentum tensor is

$$T = g^{\mu\nu}T_{\mu\nu} = \rho c^2 m^{\mu\nu}\tau_\mu\tau_\nu + \mathrm{O}(c^1) = -\rho c^2 + \mathrm{O}(c^1), \tag{3.102a}$$

implying

$$T g_{\mu\nu} = \rho c^4 \tau_\mu\tau_\nu + \mathrm{O}(c^3). \tag{3.102b}$$

Inserting (3.101c) and (3.102b) into the trace-reversed Einstein equation (3.100b), we obtain

$$\overset{g}{R}_{\mu\nu} = 4\pi G\rho\tau_\mu\tau_\nu + \mathrm{O}(c^{-1}). \tag{3.103}$$

Thus, we obtain the Newton–Cartan field equation as the formal $c \to \infty$ limit of the Einstein equation. $\triangle$

Theorem 3.53 *Assuming expansions of the Lorentzian metric and inverse metric as in Lemma 3.48, regularity of the Lorentzian Levi-Civita connection as $c \to \infty$, and an expansion of the energy–momentum tensor as in (3.101), Newton–Cartan gravity arises as the formal $c \to \infty$ limit of general relativity.*

Proof In Lemma 3.48 and Theorem 3.50, we have seen how the formal $c \to \infty$ limit of a Lorentzian spacetime gives rise to a Newtonian spacetime, including the notions of (ideal, i.e. 'metric') time and space measurements. In Construction 3.52, we have shown how the Einstein equation gives rise to the Newton–Cartan field equation. Finally, since the Levi-Civita connection of the Lorentzian spacetime expands to the Newtonian connection plus higher-order terms, Lorentzian geodesics in the limit go over to Newtonian geodesics, i.e. test particle worldlines go over to test particle worldlines. Thus, we have recovered our axioms for Newton–Cartan gravity. $\square$

Example 3.54 Consider the Schwarzschild metric

$$g = -\left(1 - \frac{2GM}{c^2 r}\right)c^2 \mathrm{d}t^2 + \left(1 - \frac{2GM}{c^2 r}\right)^{-1}\mathrm{d}r^2 + r^2(\mathrm{d}\theta^2 + \sin^2\theta\,\mathrm{d}\varphi^2),$$

$$(3.104a)$$

with inverse metric given by

$$g^{-1} = -\left(1 - \frac{2GM}{c^2 r}\right)^{-1}c^{-2}\partial_t \otimes \partial_t + \left(1 - \frac{2GM}{c^2 r}\right)\partial_r \otimes \partial_r$$
$$+ r^{-2}(\partial_\theta \otimes \partial_\theta + \sin^{-2}\theta\,\partial_\varphi \otimes \partial_\varphi).$$

$$(3.104b)$$

Expanding those in powers of c^{-1} (using the geometric series for the inversion of power series), we obtain

$$g = \underbrace{-c^2\mathrm{d}t^2}_{\stackrel{!}{=}-c^2\tau\otimes\tau} + \underbrace{\frac{2GM}{r}\mathrm{d}t^2 + \mathrm{d}r^2 + r^2(\mathrm{d}\theta^2 + \sin^2\theta\,\mathrm{d}\varphi^2)}_{=g^{(0)}} + \mathrm{O}(c^{-2}), \qquad (3.105a)$$

$$g^{-1} = \underbrace{\partial_r \otimes \partial_r + r^{-2}(\partial_\theta \otimes \partial_\theta + \sin^{-2}\theta\,\partial_\varphi \otimes \partial_\varphi)}_{=h} + c^{-2}\underbrace{\left(-\partial_t \otimes \partial_t - \frac{2GM}{r}\partial_r \otimes \partial_r\right)}_{=m}$$
$$+ \mathrm{O}(c^{-4}). \qquad (3.105b)$$

We thus see that we may take $\tau = \mathrm{d}t$, and can read off the expansion coefficients $h, g^{(0)}, m$ from Lemma 3.48. We see that on the spatial leaves of constant t, the space metric h induces just the usual Euclidean metric in spherical coordinates.

Further, we obtain $\hat{v} = -m(\tau, \cdot) = \partial_t$, giving $\hat{\phi} = -\frac{1}{2}g^{(0)}(\hat{v}, \hat{v}) = -\frac{GM}{r}$, which leads to $\underset{\hat{v}}{h} = g^{(0)} + 2\hat{\phi}\tau \otimes \tau = \mathrm{d}r^2 + r^2(\mathrm{d}\theta^2 + \sin^2\theta\,\mathrm{d}\varphi^2)$ being the Euclidean metric.

Since $\hat{v} = \partial_t$ and the coordinate components of h do not depend on t, we have $\mathcal{L}_{\hat{v}}h = 0$, i.e. $\hat{v}$ is rigid. According to Remark 3.51, it has acceleration $\hat{\alpha} = \mathrm{d}\hat{\phi}$ and is twist-free, so we may apply the Trautman recovery theorem and obtain a solution of usual Newtonian gravity with gravitational potential $\hat{\phi}$. $\triangle$

Exercise 3.55 (*The Newtonian limit of the* NUT *region*) *Taub–NUT spacetime* is an exact solution of the vacuum Einstein equation with some very peculiar properties. The Taub–NUT metric can be written as

$$g = U(c\mathrm{d}t - 4l\sin^2(\tfrac{\theta}{2})\,\mathrm{d}\varphi)^2 - \frac{1}{U}\mathrm{d}r^2 + (r^2 + l^2)(\mathrm{d}\theta^2 + \sin^2\theta\,\mathrm{d}\varphi^2), \quad (3.106a)$$

where $U = -1 + 2\big(\frac{GM}{c^2}r + l^2\big)/(r^2 + l^2)$. Here M is a mass and $l > 0$ is the so-called NUT *parameter*, θ and φ are usual spherical coordinates on S^2, and (at least a priori[27]) the range of t, r is the whole real axis $\mathbb{R}$. The inverse metric can be calculated as

$$g^{-1} = \left(\frac{1}{U} + \frac{4l^2\tan^2(\tfrac{\theta}{2})}{r^2 + l^2}\right)c^{-2}\partial_t \otimes \partial_t + \frac{l/\cos^2(\tfrac{\theta}{2})}{r^2 + l^2}c^{-1}(\partial_t \otimes \partial_\varphi + \partial_\varphi \otimes \partial_t)$$

$$- U\,\partial_r \otimes \partial_r + \frac{1}{r^2 + l^2}(\partial_\theta \otimes \partial_\theta + \sin^{-2}\theta\,\partial_\varphi \otimes \partial_\varphi). \quad (3.106b)$$

The NUT *regions* are the two regions in which $U < 0$, i.e. in which t is a timelike and r a spacelike coordinate. They are the regions $r < r_-$ and $r > r_+$ where $r_\pm = \frac{GM}{c^2} \pm \sqrt{\frac{G^2M^2}{c^4} + l^2}$.

Parametrising $l = \frac{J}{Mc}$, where J has the dimension of angular momentum, the metric in the positive r NUT region $r > r_+$ has the form which can be expanded in c^{-1} to give a Newton–Cartan spacetime. (Note that in the limit $c \to \infty$, we have $r_\pm \to 0$, such that we may use the whole positive real axis as range for r.) The objective of this exercise is to compute the formal $c \to \infty$ Newton–Cartan limit of the positive r NUT region and show that it does not have absolute rotation.

(a) Inserting $l = \frac{J}{Mc}$, expand the Taub–NUT metric and inverse metric as formal power series in c^{-1}, to order c^0 for g and order c^{-2} for g^{-1}. Comparing to the expansions from Lemma 3.48, read off the tensor fields τ, h, m, and $g^{(0)}$.
Hint: Use the geometric series. You should obtain $\tau = \mathrm{d}t$, and h should just be the inverse Euclidean metric on $\mathbb{R}^3$ in spherical coordinates.

(b) Compute the unit timelike vector field $\hat{v} = -m(\tau, \cdot)$. Show that it is not rigid.
Hint: You should obtain $\hat{v} = \partial_t - \frac{J/\cos^2(\tfrac{\theta}{2})}{Mr^2}\partial_\varphi$. For non-rigidity, you can argue that $\mathcal{L}_{\hat{v}}h$ is non-zero without completely computing it.

(c) Compute $\hat{\phi} = -\frac{1}{2}g^{(0)}(\hat{v}, \hat{v})$, and then the covariant space metric with respect to $\hat{v}$ as $\underset{\hat{v}}{h} = g^{(0)} + 2\hat{\phi}\tau \otimes \tau$.

---

[27] The Taub–NUT metric as given above has a singularity on the 'negative z-axis' $\theta = \pi$. The customary way to get rid of this singularity, following Misner, is to go over to a quotient space: one introduces $\psi := \frac{ct}{2l} - \varphi$, and then interprets ψ, θ, φ as Euler coordinates on S^3, s.t. ψ becomes 4π-periodic (for a very readable full treatment, see ref. [5, Chap. 12]). In this exercise, however, we will not make this periodic identification, since with this identification the Newtonian limit would not make sense.

(d) By Remark 3.51, the Newton–Coriolis form of the limiting Newtonian connection with respect to $\hat{v}$ is given by $\hat{\Omega} = \tau \wedge \mathrm{d}\hat{\phi}$. Compute this.

(e) We want to show that the limiting Newton–Cartan spacetime does not have absolute rotation, i.e. that rigid unit timelike vector fields on it have spatially non-constant twist. Our field $\hat{v}$ from above has no twist, but it is not rigid. We consider the rigid field $\tilde{v} = \partial_t$ instead. Using the transformation behaviour of the Newton–Coriolis form under a Milne boost ((Eq. 2.51c)), compute the Newton–Coriolis form $\tilde{\Omega}$ with respect to $\tilde{v}$. Comparing to the decomposition $\tilde{\Omega} = \tau \wedge \tilde{\alpha} + 2\tilde{\omega}$, read off the twist $\tilde{\omega}$. By expressing $\tilde{\omega}$ in terms of the spatial orthonormal coframe $\mathrm{e}^r = \mathrm{d}r$, $\mathrm{e}^\theta = r\mathrm{d}\theta$, $\mathrm{e}^\varphi = r\sin(\theta)\mathrm{d}\varphi$, read off the metric length of $\tilde{\omega}$ and observe that it is non-constant. $\triangle$

References

1. Dixon, W.G.: On the uniqueness of the Newtonian theory as a geometric theory of gravitation. Commun. Math. Phys. **45**(2), 167–182 (1975). https://doi.org/10.1007/BF01629247
2. Ehlers, J.: Über den Newtonschen Grenzwert der Einsteinschen Gravitationstheorie. In: Nitsch, J., Pfarr, J., Stachow, E.W. (eds.) Grundlagenprobleme der modernen Physik: Festschrift für Peter Mittelstaedt zum 50. Geburtstag, pp. 65–84. Bibliographisches Institut, Mannheim, Wien, Zürich (1981). Republished as [3]
3. Ehlers, J.: On the Newtonian limit of Einstein's theory of gravitation. Gen. Relativ. Gravit. **51**(12), 163 (2019). https://doi.org/10.1007/s10714-019-2624-0. Republication of original article [2] as 'Golden Oldie'
4. Ergen, M., Hamamci, E., Van den Bleeken, D.: Oddity in nonrelativistic, strong gravity. Eur. Phys. J. C **80**(6), 563 (2020). https://doi.org/10.1140/epjc/s10052-020-8112-6
5. Griffiths, J.B., Podolsk, J.: Exact Space-Times in Einstein's General Relativity. In: Cambridge monographs on mathematical physics. Cambridge University Press, Cambridge (2009). https://doi.org/10.1017/CBO9780511635397
6. Künzle, H.P.: Covariant Newtonian limit of Lorentz space-times. Gen. Relativ. Gravit. 7(5), 445–457 (1976). https://doi.org/10.1007/BF00766139
7. Malament, D.B.: Topics in the Foundations of General Relativity and Newtonian Gravitation Theory. In: Chicago lectures in physics. The University of Chicago Press, Chicago (2012). https://press.uchicago.edu/ucp/books/book/chicago/T/bo12893557.html. Available at the author's website under https://www.socsci.uci.edu/~dmalamen/bio/GR.pdf
8. Pirani, F.A.E.: A note on bouncing photons. Bull. Acad. Pol. Sci. Ser. Sci. Math. Astron. Phys. **13**, 239 (1965)
9. Schwartz, P.K.: Teleparallel Newton–Cartan gravity. Class. Quantum Gravity **40**(10), 105008 (2023). https://doi.org/10.1088/1361-6382/accc02
10. Schwartz, P.K., von Blanckenburg, A.L.: The Newtonian limit of orthonormal frames in metric theories of gravity. Gen. Relativ. Gravit. **57**, 166 (2025). https://doi.org/10.1007/s10714-025-03490-2
11. Synge, J.L.: Relativity: The General Theory. North-Holland Publishing Company, Amsterdam (1960)
12. Tichy, W., Flanagan, É.É.: Covariant formulation of the post-1-Newtonian approximation to general relativity. Phys. Rev. D **84**(4), 044038 (2011). https://doi.org/10.1103/PhysRevD.84.044038
13. Trautman, A.: Sur la théorie newtonienne de la gravitation. C. R. Acad. Sci. **257**, 617–620 (1963). https://gallica.bnf.fr/ark:/12148/bpt6k4007z/f639.item

Galilei Manifolds via Principal Bundles

4

Abstract

Here, we will discuss how we can describe Galilei manifolds and Galilei connections on them in terms of principal bundles.

4.1 The Galilei Group

Definition 4.1 The (orthochronous[1]) *homogeneous Galilei group* in $n + 1$ dimensions is the semidirect product

$$\mathrm{Gal} = \mathrm{O}(n) \ltimes \mathbb{R}^n, \tag{4.1a}$$

where $\mathrm{O}(n)$ acts on $\mathbb{R}^n$ in the natural way. This means that elements of Gal are pairs (R, k) with $R \in \mathrm{O}(n)$ and $k \in \mathbb{R}^n$, and the group operation is

$$(R, k)(\tilde{R}, \tilde{k}) = (R\tilde{R}, k + R\tilde{k}). \tag{4.1b}$$

Gal is a Lie group, with Lie algebra being the semidirect sum[2]

$$\mathfrak{gal} = \mathfrak{so}(n) \oplus \mathbb{R}^n. \tag{4.2a}$$

---

[1] Instead, we could also consider the full homogeneous Galilei group, including time reversal, or restrict to the proper (without reflections, but including time reversal) or proper orthochronous Galilei group (which is the connected component of the identity). However, our definition of Galilei manifolds comes with a choice of time orientation, as discussed in Remark 2.4, and without an orientation for space (or even the need for spatial orientability). Therefore, the orthochronous group is the one that naturally arises in the following. Of course, it is a matter of convention what precise definition of a Galilei manifold one chooses.

[2] Note that the Lie algebras of $\mathrm{O}(n)$ and $\mathrm{SO}(n)$ coincide, the latter group being the connected component of the identity of the former.

This means that as a vector space, $\mathfrak{gal}$ is the direct sum of $\mathfrak{so}(n)$ and $\mathbb{R}^n$, while the Lie bracket is given by

$$[(X, k), (\tilde{X}, \tilde{k})] = ([X, \tilde{X}], X\tilde{k} - \tilde{X}k). \tag{4.2b}$$

The homogeneous Galilei group acts faithfully on $\mathbb{R}^{n+1}$ via

$$(R, k)(y^t, y^a) = (y^t, R^a{}_b y^b + y^t k^a), \tag{4.3}$$

where $y^t \in \mathbb{R}$ and $(y^a) \in \mathbb{R}^n$. In this way, it can be viewed as a subgroup of $GL(n + 1)$. ❀

Notation 4.2 When understanding $\mathbb{R}^{n+1}$ as the space on which the Galilei group acts according to (4.3), we use an index notation similar to that for Galilei bases introduced in Notation 2.14: the first component of elements of $\mathbb{R}^{n+1}$ is labelled by an index t, the other components are labelled by lowercase Latin letters from the beginning of the alphabet, running from 1 to n. Capital Latin indices run through the full range $\{t, 1, \ldots, n\}$. Hence, we may write an element of $\mathbb{R}^{n+1}$ as $(y^A) = (y^t, y^a)$. ❀

Interpretation 4.3 Interpreting y^t as the time and y^a as the position of an event in the standard coordinate formulation of Newtonian mechanics, we see that under the action (4.3) the $O(n)$ part of Gal corresponds to improper rotations of space, while the $\mathbb{R}^n$ part corresponds to Galilei boosts. △

Remark 4.4 Writing the action (4.3) of Gal on $\mathbb{R}^{n+1}$ in matrix form, we see that when viewing Gal as a subgroup of $GL(n + 1)$, the matrix corresponding to a Galilei group element $(R, k) \in \mathsf{Gal}$ is

$$\begin{pmatrix} 1 & 0 \\ k^a & R^a{}_b \end{pmatrix} \in GL(n + 1). \tag{4.4a}$$

Therefore, the condition for a matrix $A \in GL(n + 1)$ to be an element of Gal can be written in components as

$$A^t{}_t = 1, \quad A^t{}_a = 0, \quad (A^a{}_b) \in O(n). \tag{4.4b}$$

Viewing the homogeneous Galilei Lie algebra $\mathfrak{gal}$ as a subalgebra of the matrix Lie algebra $\mathfrak{gl}(n + 1)$ of all $(n + 1) \times (n + 1)$ matrices, the abstract Lie algebra element $(X, k) \in \mathfrak{gal} = \mathfrak{so}(n) \oplus \mathbb{R}^n$ is represented by the matrix

$$\begin{pmatrix} 0 & 0 \\ k^a & X^a{}_b \end{pmatrix} \in \mathfrak{gl}(n + 1). \tag{4.5a}$$

In components, the condition for a matrix $Y \in \mathfrak{gl}(n + 1)$ to lie in $\mathfrak{gal}$ therefore reads

$$Y^t{}_A = 0, \quad (Y^a{}_b) \in \mathfrak{so}(n). \tag{4.5b}$$

△

4.2 Galilei Structures

Lemma 4.5 *Let (M, τ, h) be a Galilei manifold and (e_A) a Galilei basis at $p \in M$. Another basis $(\tilde{e}_A)$ of $T_p M$ is a Galilei basis if and only if the basis change matrix $A \in \mathsf{GL}(n+1)$, defined by $\tilde{e}_A = A^B{}_A e_B$, is an element of the homogeneous Galilei group Gal (understood as a subgroup of $\mathsf{GL}(n+1)$ as in Remark 4.4).*

Proof We first recall that if two bases of a vector space are related by $\tilde{e}_A = A^B{}_A e_B$, then the dual bases are related by $\tilde{e}^A = (A^{-1})^A{}_B e^B$. Of course, we can also invert the original relation to $e_A = (A^{-1})^B{}_A \tilde{e}_B$. The second basis $(\tilde{e}_A)$ of $T_p M$ being a Galilei basis means that

$$\tilde{e}^t = \tau|_p = e^t, \quad \delta^{ab}\tilde{e}_a \otimes \tilde{e}_b = h|_p = \delta^{ab} e_a \otimes e_b \,. \tag{4.6}$$

Expressing $\tilde{e}^t$ in terms of the e^A in the first equation and the e_a in terms of the $\tilde{e}_A$ in the second one, these are equivalent to

$$(A^{-1})^t{}_t e^t + (A^{-1})^t{}_a e^a = e^t, \tag{4.7a}$$

i.e.

$$(A^{-1})^t{}_t = 1, \quad (A^{-1})^t{}_a = 0, \tag{4.7b}$$

and

$$\begin{aligned}
\delta^{ab}\tilde{e}_a \otimes \tilde{e}_b &= \delta^{ab}\left((A^{-1})^t{}_a\tilde{e}_t + (A^{-1})^c{}_a\tilde{e}_c\right) \otimes \left((A^{-1})^t{}_b\tilde{e}_t + (A^{-1})^d{}_b\tilde{e}_d\right) \\
&= (A^{-1})^c{}_a\delta^{ab}(A^{-1})^d{}_b\tilde{e}_c \otimes \tilde{e}_d \\
&= (A^{-1})^a{}_c\delta^{cd}(A^{-1})^b{}_d\tilde{e}_a \otimes \tilde{e}_b \,,
\end{aligned} \tag{4.7c}$$

i.e.

$$\delta^{ab} = (A^{-1})^a{}_c\delta^{cd}(A^{-1})^b{}_d \iff ((A^{-1})^a{}_b) \in \mathsf{O}(n). \tag{4.7d}$$

Comparing to Remark 4.4, we see that this means that A^{-1} is an element of Gal, which is of course equivalent to A being an element of Gal. $\qquad\square$

We can use this observation to combine all Galilei bases into a principal bundle:

Definition 4.6 Let (M, τ, h) be a Galilei manifold.

(i) A *(local) Galilei frame* on (M, τ, h) is a local frame (e_A) of vector fields on an open set $U \subset M$ such that for each $p \in U$, the basis $(e_A|_p)$ of $T_p M$ is a Galilei basis.

(ii) The *Galilei frame bundle* of (M, τ, h) is the set of all Galilei bases at all points of M, denoted by $G(M)$.[3] A priori, this is simply a subset of the linear frame bundle $F(M)$. However, combined with the observation that smooth local Galilei frames exist around any point[4], Lemma 4.5 shows that $G(M)$ is a reduction of the structure group of $F(M)$ from $\mathsf{GL}(n + 1)$ to Gal (viewed as a subgroup of $\mathsf{GL}(n + 1)$ as in Remark 4.4). By construction, the Galilei frames on (M, τ, h) are precisely the local sections of the Galilei frame bundle $G(M)$.

Explicitly spelling out the action of Gal on $G(M)$, it reads

$$(e_t|_p, e_a|_p) \cdot (R, k) = (e_t|_p + k^b e_b|_p, e_b|_p R^b{}_a) \tag{4.8}$$

for $(e_A|_p) \in G(M)$ and $(R, k) \in \mathsf{Gal}$.   ✿

Notation 4.7 We will usually denote the unit timelike field of a Galilei frame as $e_t =: v$, conforming to our previous notation for (local) choices of unit timelike future-directed vector fields.   ✿

Construction 4.8 We can reverse the construction of the Galilei frame bundle from τ and h as in Definition 4.6 in the following sense: for an $(n + 1)$-dimensional manifold M, given a reduction $G(M)$ of the structure group of the linear frame bundle $F(M)$ to Gal, there are a unique clock form τ and space metric h making M into a Galilei manifold such that $G(M)$ is the Galilei frame bundle of (M, τ, h). Namely, given any local section $(e_A) \in \Gamma(U, G(M)) \subset \Gamma(U, F(M))$, we can define h and τ on U by $h := \delta^{ab} e_a \otimes e_b$ and $\tau := e^t$, where (e^A) is the dual frame to (e_A). Smoothness of τ and h thus defined is immediate. That they indeed define a Galilei manifold follows directly by linear independence of the e_a and by definition of the dual basis. Finally, τ and h are well-defined, i.e. independent of the choice of local section, due to Gal-invariance of $G(M)$: any other local section $(\tilde{e}_A)$ is related to (e_A) by a Galilei transformation at each point, so by Lemma 4.5 $(\tilde{e}_A|_p)$ is also a Galilei basis with respect to τ and h, i.e. $\tau = \tilde{e}^t$ and $h = \delta^{ab} \tilde{e}_a \otimes \tilde{e}_b$.   △

Put differently, the data τ, h that make a manifold into a Galilei manifold are equivalent to a choice of which linear frames are Galilei frames.

Definition 4.9 A *Galilei structure* on an $(n + 1)$-dimensional smooth manifold M is a reduction $G(M)$ of the structure group of the linear frame bundle $F(M)$ from $\mathsf{GL}(n + 1)$ to the homogeneous Galilei group Gal.   ✿

---

[3] The Galilei frame bundle depends on τ and h, but we will not acknowledge this in our notation.

[4] A smooth local Galilei frame around $p \in M$ may be constructed as follows. We choose *any* smooth frame of vector fields $(\tilde{e}_A)$ on some neighbourhood of p such that $\tilde{e}_t =: e_t$ is a unit timelike future-directed field. Then we project the other fields $\tilde{e}_a$ onto space along $\tilde{e}_t$ and apply the Gram–Schmidt orthonormalisation process with respect to ${}^{(n)}h$ to them. Due to smoothness of τ and ${}^{(n)}h$, the resulting fields e_a are smooth. By construction, (e_A) will then at each point be a Galilei basis.

Notation 4.10 Since the Galilei frame bundle $G(M)$ is a reduction of the structure group of the linear frame bundle $F(M)$, we obtain the tangent bundle TM as an associated vector bundle,

$$TM \cong G(M) \times_{\mathsf{Gal}} \mathbb{R}^{n+1}. \tag{4.9}$$

The isomorphism is given by the canonical solder form $\theta \in \Omega^1(M, G(M) \times_{\mathsf{Gal}} \mathbb{R}^{n+1}) = \mathrm{Hom}(TM, G(M) \times_{\mathsf{Gal}} \mathbb{R}^{n+1})$,

$$\theta_p(v^A \mathrm{e}_A|_p) = [(\mathrm{e}_A|_p), (v^A)]. \tag{4.10}$$

Using the representations of Gal induced by tensor representations of $\mathsf{GL}(n+1)$, we further obtain all tensor bundles of M as vector bundles associated to the Galilei frame bundle. Concretely, understanding any tensor on M as an element of the respective associated bundle, its representative with respect to a chosen local Galilei frame is simply given by the tensor's components with respect to this frame.

When a Galilei frame (e_A) is chosen, we denote tensor components with respect to this frame by the same kind of indices $(A) = (t, a)$ that we use for the frame vectors. For example, we may locally write a vector field $X \in \Gamma(TM)$ as

$$X = X^A \mathrm{e}_A \text{ with } X^A = \mathrm{e}^A(X) = \mathrm{e}^A_\mu X^\mu, \tag{4.11}$$

or a one-form $\alpha \in \Omega^1(M)$ as

$$\alpha = \alpha_A \mathrm{e}^A \text{ with } \alpha_A = \alpha(\mathrm{e}_A) = \mathrm{e}^\mu_A \alpha_\mu. \tag{4.12}$$

So from a purely notational point of view, we can use the components e^μ_A of a Galilei frame and e^A_μ of the dual frame to 'convert' tensor indices from 'coordinate indices' μ to 'Galilei frame indices' A and back. ✿

Construction 4.11 We now want to consider principal connections ω on the Galilei frame bundle $G(M)$ of a Galilei manifold (M, τ, h).[5] The connection form ω is a one-form on $G(M)$ with values in the Galilei Lie algebra $\mathfrak{gal} = \mathfrak{so}(n) \oplus \mathbb{R}^n$, i.e. it can be decomposed as

$$\omega = (\omega^a{}_b, \varpi^a) \in \Omega^1(G(M), \mathfrak{so}(n) \oplus \mathbb{R}^n) \tag{4.13}$$

with an $\mathfrak{so}(n)$-valued part $(\omega^a{}_b)$ and an $\mathbb{R}^n$-valued part (ϖ^a). The local connection form[6] ω with respect to a Galilei frame (e_A) on $U \subset M$, i.e. the pullback of the

---

[5] Some of the discussion in this construction is in parallel to parts of my (the author's) article [1].

[6] Note that we denote principal connections by the same letter ω that we earlier used to refer to the twist of a vector field. The meaning should always be clear from context.

connection form $\omega \in \Omega^1(G(M), \mathfrak{gal})$ along (e_A), is then a $\mathfrak{gal}$-valued one-form on U, which we can similarly decompose as

$$\omega = (\omega^a{}_b, \varpi^a) \in \Omega^1(U, \mathfrak{so}(n) \oplus \mathbb{R}^n) \tag{4.14}$$

Via Remark 4.4, we may also understand this as a matrix-valued one-form $(\omega^A{}_B) \in \Omega^1(U, \mathfrak{gl}(n+1))$ with

$$\omega^t{}_A = 0, \quad (\omega^a{}_b) \in \Omega^1(U, \mathfrak{so}(n)), \quad \omega^a{}_t = \varpi^a, \tag{4.15a}$$

i.e.

$$(\omega^A{}_B) = \begin{pmatrix} 0 & 0 \\ \varpi^a & \omega^a{}_b \end{pmatrix} \in \Omega^1(U, \mathfrak{gl}(n+1)). \tag{4.15b}$$

We now consider the covariant derivative operator ∇ induced by ω on the tangent bundle $TM \cong G(M) \times_{\mathsf{Gal}} \mathbb{R}^{n+1}$ and the tensor bundles constructed from it. As for any principal connection on a reduction of the structure group of the linear frame bundle, the action of the induced linear connection ∇ may locally be expressed in terms of the matrix-valued local connection form $(\omega^A{}_B)$: it acts on the frame fields e_A and the dual frame fields e^A according to

$$\nabla e_B = \omega^A{}_B \otimes e_A, \quad \nabla e^A = -\omega^A{}_B \otimes e^B, \tag{4.16a}$$

and for any tensor field with frame components $X^{A_1...A_p}_{B_1...B_q}$, its covariant derivative—viewed as a $(\bigotimes^p TM \otimes \bigotimes^q T^*M)$-valued one-form—has frame components

$$(\nabla X)^{A_1...A_p}_{B_1...B_q} = \mathrm{d}X^{A_1...A_p}_{B_1...B_q} + \omega^{A_1}{}_C\, X^{CA_2...A_p}_{B_1...B_q} + \cdots - \omega^D{}_{B_1}\, X^{A_1...A_p}_{DB_2...B_q} - \cdots . \tag{4.16b}$$

We can use this to compute $\nabla\tau$ and ∇h: for a Galilei frame (e_A), we have $\tau = e^t$ and $h = \delta^{ab} e_a \otimes e_b$, i.e. the frame components of τ and h are

$$\tau_A = \delta^t_A, \quad h^{AB} = \delta^{ab}\delta^A_a\delta^B_b . \tag{4.17}$$

Thus, $\nabla\tau$ and ∇h—viewed as T^*M- and $(\bigvee^2 TM)$-valued one-forms, respectively—have frame components

$$(\nabla\tau)_A = \mathrm{d}\tau_A - \omega^B{}_A\tau_B = 0 - \omega^B{}_A\delta^t_B = -\omega^t{}_A \tag{4.18}$$

and

$$\begin{aligned}
(\nabla h)^{AB} &= \mathrm{d}h^{AB} + \omega^A{}_c h^{CB} + \omega^B{}_c h^{AC} \\
&= 0 + \omega^A{}_c \delta^{cb}\delta^B_b + \omega^B{}_c \delta^{ac}\delta^A_a \\
&= 2\omega^{(A}{}_c \delta^{B)}_b \delta^{bc} .
\end{aligned} \tag{4.19a}$$

Separating the latter into timelike and spacelike components, we have

$$(\nabla h)^{tt} = 0 , \quad (\nabla h)^{ta} = (\nabla h)^{at} = \omega^t{}_b \delta^{ab} , \quad (\nabla h)^{ab} = 2\omega^{(a}{}_c \delta^{b)c} . \quad (4.19\text{b})$$

Now combining (4.18) and (4.19b) with (4.15a, b), we obtain that the induced ∇ is a Galilei connection on (M, τ, h).

Conversely, given a Galilei connection ∇ on (M, τ, h), (4.18) and (4.19b) show that its local connection form $(\omega^A{}_B) \in \Omega^1(U, \mathfrak{gl}(n+1))$ with respect to a Galilei frame (e_A) on U satisfies (4.15a, b), i.e. takes values in $\mathfrak{gal}$. Therefore, ∇ gives rise to a principal connection ω on the Galilei frame bundle $G(M)$ by which it is induced. $\triangle$

Thus we have seen that Galilei connections on a Galilei manifold are essentially the same as principal connections on the Galilei frame bundle:

Proposition 4.12 *Let (M, τ, h) be a Galilei manifold with Galilei frame bundle $G(M)$.*

(i) Given a principal connection ω on $G(M)$, the induced covariant derivative operator ∇ on the tangent bundle $TM \cong G(M) \times_{\mathrm{Gal}} \mathbb{R}^{n+1}$ is a Galilei connection on (M, τ, h).

(ii) Conversely, given a Galilei connection ∇ on (M, τ, h), it is induced by a (unique) principal connection ω on $G(M)$. $\square$

Remark 4.13 From this correspondence we obtain yet another way to show that the temporal torsion of a Galilei connection on (M, τ, h) is $\mathrm{d}\tau$ (Proposition 2.20 (i)): by Cartan's first structure equation, the torsion—understood as a TM-valued two-form—is the exterior covariant derivative of the canonical solder form. Hence, its temporal component is given by

$$T^t = (\mathrm{d}^\omega \theta)^t = \mathrm{d}\theta^t + \omega^t{}_B \wedge \theta^B = \mathrm{d}\tau \quad (4.20)$$

where we used that $\theta^t = e^t = \tau$ and that $\omega^t{}_B = 0$ by (4.15a).

The general properties of the curvature tensor of a Galilei manifold (Proposition 2.20 (ii)), namely the vanishing of the temporal curvature and the antisymmetry of the curvature tensor with second index raised with h, also have a simple interpretation in terms of the principal bundle formalism: in terms of the local curvature form $R \in \Omega^2(U, \mathfrak{gl}(n+1))$ of ω, the curvature tensor is

$$\frac{1}{2} R^\mu{}_{\nu\rho\sigma} \, \mathrm{d}x^\rho \wedge \mathrm{d}x^\sigma = e^\mu_A e^B_\nu R^A{}_B , \quad (4.21)$$

and so these properties of $R^\mu{}_{\nu\rho\sigma}$ are equivalent (via Remark 4.4) to the local curvature form R taking values in $\mathfrak{gal} \subset \mathfrak{gl}(n+1)$. $\triangle$

Proposition 4.14 *Let (M, τ, h) be a Galilei manifold with Galilei frame bundle $G(M)$, and ∇ a Galilei connection on (M, τ, h) with corresponding principal connection ω on $G(M)$. Let $(e_A) = (v, e_a)$ be a local Galilei frame.*

Then the Newton–Coriolis form of ∇ with respect to v can be written in terms of the 'boost part' of the local connection form and the dual frame as

$$\Omega = \delta_{ab}\varpi^a \wedge e^b, \tag{4.22a}$$

i.e. in components

$$\Omega_{\mu\nu} = \delta_{ab}2\varpi_{[\mu}{}^a e_{\nu]}^b . \tag{4.22b}$$

Writing $\varpi_\mu{}^\nu := \varpi_\mu{}^a e_a^\nu$, this may be expressed as

$$\Omega_{\mu\nu} = 2\varpi_{[\mu\nu]} , \tag{4.23}$$

i.e. in some sense the Newton–Coriolis form is the antisymmetrisation of the boost part of the local connection form.

Proof The definition of the Newton–Coriolis form reads $\Omega_{\mu\nu} = 2(\nabla_{[\mu}v^\rho)h_{\nu]\rho}$. Expressing the covariant space metric in terms of the Galilei frame (Proposition 2.18 (ii)) and using the definition of the local connection form, we obtain

$$\Omega_{\mu\nu} = 2(\nabla_{[\mu}v^\rho)h_{\nu]\rho} = 2(\nabla_{[\mu|}e_t^\rho)\delta_{ab}e_{|\nu]}^a e_\rho^b$$
$$= 2\omega_{[\mu|}{}^A{}_t e_A^\rho \delta_{ab}e_{|\nu]}^a e_\rho^b = 2\varpi_{[\mu|}{}^c e_c^\rho \delta_{ab}e_{|\nu]}^a e_\rho^b . \tag{4.24}$$

Combining $e_c^\rho e_\rho^b = \delta_c^b$, this gives

$$\Omega_{\mu\nu} = 2\varpi_{[\mu}{}^c e_{\nu]}^a \delta_{ab}\delta_c^b = 2\varpi_{[\mu}{}^b e_{\nu]}^a \delta_{ab} . \tag{4.25}$$

Instead using $\varpi_\mu{}^c e_c^\rho = \varpi_\mu{}^\rho$, we get

$$\Omega_{\mu\nu} = 2\varpi_{[\mu}{}^\rho e_{\nu]}^a \delta_{ab}e_\rho^b = 2\varpi_{[\mu}{}^\rho h_{\nu]\rho} = 2\varpi_{[\mu\nu]} . \tag{4.26}$$

$\square$

Definition 4.15 Any change of local Galilei frame has the form

$$(v, e_a) \mapsto (\tilde{v}, \tilde{e}_a) = (v, e_a) \cdot (R, k)^{-1} \tag{4.27a}$$

for a local **Gal**-valued function (R, k). This is called a *local Galilei transformation*. Spelling out the action via (4.8) and using $(R, k)^{-1} = (R^{-1}, -R^{-1}k)$, it reads

$$(v, e_a) \mapsto (\tilde{v}, \tilde{e}_a) = \left(v - e_b(R^{-1})^b{}_a k^a, e_b(R^{-1})^b{}_a\right). \tag{4.27b}$$

This defines a left action on Galilei frames by the group of local **Gal**-valued functions.   ✤

Note that local Galilei *boosts*, i.e. local Galilei transformations with $(R, k) = (\mathbb{1}, k)$ for some $\mathbb{R}^n$-valued function k, have the form

$$(v, \mathrm{e}_a) \mapsto (\tilde{v}, \tilde{\mathrm{e}}_a) = (v - \mathrm{e}_a k^a, \mathrm{e}_a). \tag{4.28}$$

Comparing this to (2.50), we see that this is a Milne boost of the unit timelike vector field v with parameter field $w = k^a \mathrm{e}_a$—*Milne boosts are local Galilei boosts!*

Under local Galilei transformations of the frame, the dual frame and the local connection forms transform as follows:

Proposition 4.16

(i) Under a local Galilei transformation (4.27a, b) which is purely rotational, i.e. with $k = 0$, the dual frame and local connection forms transform as

$$(\tau, \mathrm{e}^a) \mapsto (\tau, R^a{}_b \mathrm{e}^b), \tag{4.29a}$$

$$\omega^a{}_b \mapsto R^a{}_c \omega^c{}_d (R^{-1})^d{}_b + R^a{}_c \mathrm{d}(R^{-1})^c{}_b , \tag{4.29b}$$

$$\varpi^a \mapsto R^a{}_b \varpi^b. \tag{4.29c}$$

(ii) Under a local Galilei boost with parameter k, the dual frame and local connection forms transform as

$$(\tau, \mathrm{e}^a) \mapsto (\tau, \mathrm{e}^a + k^a \tau), \tag{4.30a}$$

$$\omega^a{}_b \mapsto \omega^a{}_b , \tag{4.30b}$$

$$\varpi^a \mapsto \varpi^a - \mathrm{d}k^a - \omega^a{}_b k^b. \tag{4.30c}$$

Proof Exercise 4.17. $\qquad\qquad\qquad\qquad\qquad\qquad\qquad\qquad\qquad\qquad\qquad\square$

Exercise 4.17 Let $(\mathrm{e}_A) = (v, \mathrm{e}_a)$ be a local Galilei frame on a Galilei manifold (M, τ, h). In this exercise, we are going to prove Proposition 4.16.

(a) Determine how the dual frame $(\mathrm{e}^A) = (\tau, \mathrm{e}^a)$ and the local connection form $(\omega^a{}_b, \varpi^a)$ of a Galilei connection change under a purely rotational local Galilei transformation, i.e. a local Galilei transformation with $k = 0$.
   Hint: First write the frame change in matrix form. Then use that for a basis change $\tilde{\mathrm{e}}_A = A^B{}_A \mathrm{e}_B$, the dual basis changes as $\tilde{\mathrm{e}}^A = (A^{-1})^A{}_B \mathrm{e}^B$. For the connection form, use that for a connection on a principal bundle whose structure group is a matrix Lie group, local connection forms in general transform under change of local section according to

$$(\sigma \cdot g^{-1})^* \omega = g \cdot (\sigma^* \omega) \cdot g^{-1} + g \cdot \mathrm{d}(g^{-1}). \tag{4.31}$$

You should obtain (4.29a, b, c).

(b) Determine how $(e^A) = (\tau, e^a)$ and $(\omega^a{}_b, \varpi^a)$ transform under a local Galilei boost, i.e. a local Galilei transformation with $R = \mathbb{1}$.

Hint: Proceed as for the rotations. You should obtain (4.30a, b, c). $\triangle$

Using the result (4.30a, b, c), we can now actually *derive* the transformation behaviour of the covariant space metric $\underset{v}{h}$ and the Newton–Coriolis form Ω under Milne boosts, which was just stated out of nowhere in the statement of Proposition 2.37:

Exercise 4.18 Using the transformation behaviour of (e^A) and ϖ^a as given in (4.30a, b, c), compute how the covariant space metric $\underset{v}{h}$ and the Newton–Coriolis form Ω with respect to v transform under the Milne boost $v \mapsto \tilde{v} = v - k^a e_a$ parametrised by (k^a). Show that you obtain the behaviour as stated in Proposition 2.37.

Hint: Express $\underset{v}{h}$ and Ω in terms of e^A and ϖ according to (2.14) and (4.22a), then apply the transformation (4.30a, b, c). To compare to Proposition 2.37, note that for the vector field $w = k^a e_a$, the corresponding one-form is $w^\flat = \underset{v}{h}(w, \cdot) = \delta_{ab} k^a e^b$ (why?). Compute $\mathrm{d}w^\flat$ from this expression, and re-express $\mathrm{d}e^a$ in terms of the torsion via Cartan's first structure equation. $\triangle$

Reference

1. Schwartz, P.K.: The classification of general affine connections in Newton–Cartan geometry: towards metric-affine Newton–Cartan gravity. Class. Quantum Gravity **42**(1), 015010 (2025). https://doi.org/10.1088/1361-6382/ad922f

Bargmann Forms

5

Abstract

In this chapter, we will use the principal bundle description of Galilei manifolds to introduce the notion of so-called Bargmann forms. These allow for a *global* formulation of the classification theorem for Galilei connections, i.e. one that is independent of a choice of unit timelike vector field. We will also see how the formalism of Bargmann forms enables a description of the coupling of massive matter to Newton–Cartan gravity via variational principles, and how it allows for a more general and computationally easier description of the limit from GR to Newton–Cartan gravity than that we encountered before. Finally, we will discuss how the quotient of a Lorentzian manifold by a lightlike symmetry naturally carries the structure of a Galilei manifold.

5.1 The Bargmann Group

Definition 5.1 The *Bargmann group* in $n + 1$ dimensions is the semidirect product

$$\mathsf{Barg} = \mathsf{Gal} \ltimes_\rho (\mathbb{R}^{n+1} \times \mathbb{R}), \tag{5.1a}$$

where the homomorphism $\rho \colon \mathsf{Gal} \to \mathsf{Aut}(\mathbb{R}^{n+1} \times \mathbb{R})$ is given by

$$\rho_{(R,k)}(y^A, \beta) = \left(y^t, R^a{}_b y^b + y^t k^a, \beta + \frac{1}{2}|k|^2 y^t + k_a R^a{}_b y^b \right) \tag{5.1b}$$

with $|k|^2 = \delta_{ab} k^a k^b$ and $k_a = \delta_{ab} k^b$.[1] ✽

---

[1] Often, discussions of the Bargmann group replace the $\mathbb{R}$ factor by $\mathsf{U}(1)$. We work with $\mathbb{R}$ instead simply for the sake of notational convenience: this way, we do not have to write imaginary units or complex exponentials. For our discussions, the global topological structure of the Bargmann group will be irrelevant.

© The Author(s), under exclusive license to Springer Nature Switzerland AG 2026   85
P. K. Schwartz, *Newton–Cartan Gravity*, Lecture Notes in Physics 1044,
https://doi.org/10.1007/978-3-032-03967-5_5

Exercise 5.2 Show that the induced homomorphisms $\dot{\rho}\colon \mathrm{Gal} \to \mathrm{Aut}(\mathbb{R}^{n+1} \oplus \mathbb{R})$ and $\dot{\rho}'\colon \mathfrak{gal} \to \mathrm{Der}(\mathbb{R}^{n+1} \oplus \mathbb{R})$ are given by

$$\dot{\rho}_{(R,k)}(y^A, \beta) = \left((R,k)^A{}_B y^B, \beta + \frac{1}{2}|k|^2 y^t + k_a R^a{}_b y^b \right), \tag{5.2}$$

$$\dot{\rho}'_{(X,k)}(y^A, \beta) = \left((X,k)^A{}_B y^B, k_a y^a \right), \tag{5.3}$$

where $(R,k)^A{}_B$ denotes the components of $(R,k) \in \mathrm{Gal}$ understood as a matrix in $\mathrm{GL}(n+1)$ via Remark 4.4, and similarly $(X,k)^A{}_B$ denotes the components of $(X,k) \in \mathfrak{gal}$ understood as a matrix in $\mathfrak{gl}(n+1)$.

Hint: What does the general definition from the theory of semidirect products translate to in the concrete case at hand? (It may be found in the appendix in Construction A.5.) △

The Bargmann algebra $\mathfrak{barg} = \mathfrak{gal} \oplus (\mathbb{R}^{n+1} \oplus \mathbb{R})$ is a one-dimensional central extension of the inhomogeneous Galilei algebra $\mathfrak{igal} = \mathfrak{gal} \oplus \mathbb{R}^{n+1}$, i.e. we have a short exact sequence

$$0 \to \mathbb{R} \to \mathfrak{barg} \to \mathfrak{igal} \to 0 \tag{5.4}$$

of Lie algebras where the image of $\mathbb{R}$ in $\mathfrak{barg}$ lies in the centre (i.e. commutes with all elements). In fact, up to isomorphism and choice of a constant prefactor in the second component of (5.3), the Bargmann algebra is, for $n \neq 2$, the *unique* one-dimensional non-trivial central extension of the inhomogeneous Galilei algebra.[2]

5.2 Foundations of Bargmann Forms

Construction 5.3 Denote by $\rho\colon \mathrm{Gal} \to \mathrm{Aut}(\mathbb{R}^{n+1} \times \mathbb{R})$ the group homomorphism from the definition of the Bargmann group (Definition 5.1). Let $\dot{\rho}\colon \mathrm{Gal} \to \mathrm{Aut}(\mathbb{R}^{n+1} \oplus \mathbb{R})$ be the induced representation of Gal on $\mathbb{R}^{n+1} \oplus \mathbb{R}$ by Lie algebra automorphisms (which are the same as invertible linear maps, since $\mathbb{R}^{n+1} \oplus \mathbb{R}$ is an abelian Lie algebra).

From the explicit form (5.2) of $\dot{\rho}$, we see that the $\mathbb{R}$ component β of the argument does not enter the $\mathbb{R}^{n+1}$ component of the result—i.e. the second direct summand $\mathbb{R} \subset \mathbb{R}^{n+1} \oplus \mathbb{R}$ is a $\dot{\rho}$-invariant subspace. Thus we may form the quotient representation

$$\mathrm{Gal} \to \mathrm{GL}((\mathbb{R}^{n+1} \oplus \mathbb{R})/\mathbb{R}) \cong \mathrm{GL}(n+1). \tag{5.5}$$

Here we used the natural identification $(\mathbb{R}^{n+1} \oplus \mathbb{R})/\mathbb{R} \cong \mathbb{R}^{n+1}$ of the quotient with the first direct summand. Again looking at the explicit form (5.2) of $\dot{\rho}$, we see that this quotient representation is the usual representation of Gal on $\mathbb{R}^{n+1}$ (as in Remark 4.4).

---

[2] In $n = 2$ spatial dimensions, the space of 1d central extensions of the inhomogeneous Galilei algebra is three-dimensional instead of one-dimensional.

Therefore the following definition makes sense: △

Definition 5.4 A *Bargmann form* on a Galilei manifold (M, τ, h) is a ($\mathbb{R}$-valued) one-form $a \in \Omega^1(G(M))$ on the Galilei frame bundle that together with the tensorial form $\theta \in \Omega^1_{\mathrm{Gal}}(G(M), \mathbb{R}^{n+1})$ corresponding to the canonical solder form of $G(M) \times_{\mathrm{Gal}} \mathbb{R}^{n+1}$ combines into a $\dot\rho$-tensorial form $(\theta, a) \in \Omega^1_{\dot\rho}(G(M), \mathbb{R}^{n+1} \oplus \mathbb{R})$.[3]   ❀

Construction 5.5 Let a be a Bargmann form on a Galilei manifold (M, τ, h). Given a local Galilei frame $\sigma = (\mathrm{e}_A)$ defined on an open set $U \subset M$, we consider the local representative $a = \sigma^* a \in \Omega^1(U)$ of the Bargmann form. We are now going to determine the behaviour of this representative under a local Galilei transformation $\sigma = (\mathrm{e}_A) \mapsto \tilde\sigma = (\tilde{\mathrm{e}}_A) = \sigma \cdot (R, k)^{-1}$.

Since (θ, a) is $\dot\rho$-tensorial, we have

$$\tilde\sigma^*(\theta, a) = (\sigma \cdot (R, k)^{-1})^*(\theta, a) = \dot\rho_{(R,k)} \circ (\sigma^*(\theta, a)). \tag{5.6}$$

Because θ corresponds to the canonical solder form, we have $\sigma^*\theta = (\mathrm{e}^A)$ and $\tilde\sigma^*\theta = (\tilde{\mathrm{e}}^A)$ (the dual frames). Thus, writing $\tilde a = \tilde\sigma^* a$, (5.6) becomes $(\tilde{\mathrm{e}}^A, \tilde a) = \dot\rho_{(R,k)} \circ (\mathrm{e}^A, a)$. Using the explicit form (5.2) of $\dot\rho$, this reads

$$(\tilde{\mathrm{e}}^A, \tilde a) = \left((R, k)^A{}_B \mathrm{e}^B, \, a + \frac{1}{2}|k|^2\tau + k_a R^a{}_b \mathrm{e}^b \right). \tag{5.7}$$

We may rephrase this as follows: under local *rotations* of Galilei frames $(v, \mathrm{e}_a) \mapsto (v, \mathrm{e}_b(R^{-1})^b{}_a)$, the local representative of the Bargmann form is invariant; under local Galilei *boosts* $(v, \mathrm{e}_a) \mapsto (v - \mathrm{e}_a k^a, \mathrm{e}_a)$, it transforms according to

$$a \mapsto a + k_a \mathrm{e}^a + \frac{1}{2}|k|^2\tau. \tag{5.8}$$

Conversely, this yields a local definition of a Bargmann form: given an assignment of a local one-form $a \in \Omega(U)$ to each local Galilei frame (e_A) on $U \subset M$ which is invariant under local rotations and behaves according to (5.8) under local Galilei boosts, these local forms 'patch up' to a unique Bargmann form a.   △

---

[3] In most of the modern literature on Newton–Cartan gravity, the object we call a 'Bargmann form'—or, more precisely, its local representative, see Construction 5.5—is either given no name or called the 'mass gauge field/potential' or similar. However, the latter name is misleading when describing Newtonian limits, since in this case a does not have the interpretation of a gauge potential related to the central 'mass direction' of the Bargmann algebra—see Remarks 5.20 and 5.29 for details. Therefore, we use the name 'Bargmann form' instead. In my (the author's) article [12], I used the terminology 'Bargmann structure' instead, which however (a) has been used with a different meaning in previous literature [3] and (b) is potentially confusing because of the notion of G-structures on manifolds.

Definition 5.6 Let (M, τ, h) be a Galilei manifold and $a \in \Omega^1(G(M))$ a Bargmann form on it. For a Galilei connection ∇ on (M, τ, h) with corresponding principal connection ω on $G(M)$, its *extended torsion* with respect to the Bargmann form a is the exterior covariant derivative $\mathrm{d}^\omega(\theta, a) \in \Omega^2_\rho(G(M), \mathbb{R}^{n+1} \oplus \mathbb{R})$ of the tensorial form $(\theta, a) \in \Omega^1_\rho(G(M), \mathbb{R}^{n+1} \oplus \mathbb{R})$ with respect to ω, where θ corresponds to the canonical solder form.

Decomposing the extended torsion as

$$\mathrm{d}^\omega(\theta, a) = (T, f) \in \Omega^2_\rho(G(M), \mathbb{R}^{n+1} \oplus \mathbb{R}), \tag{5.9}$$

by Cartan's first structure equation its $\mathbb{R}^{n+1}$-valued part $T \in \Omega^2_{\mathsf{Gal}}(G(M), \mathbb{R}^{n+1})$ corresponds to the torsion of ∇. Its $\mathbb{R}$-valued part $f \in \Omega^2(G(M))$ we call the *mass torsion*.[4] Note that similarly to the Bargmann form, the mass torsion itself is not tensorial; only together with T it combines into the ρ-tensorial form (T, f). ❀

The terms 'extended torsion' and 'mass torsion' were introduced in the article [7], where however they referred to the local representatives of the global objects discussed here.[5]

Proposition 5.7 *Let (M, τ, h) be a Galilei manifold and ∇ a Galilei connection on it with corresponding principal connection $\omega = (\omega^a{}_b, \varpi^a) \in \Omega^1(G(M), \mathfrak{gal}) = \Omega^1(G(M), \mathfrak{so}(n) \oplus \mathbb{R}^n)$. Let $a \in \Omega^1(G(M))$ be a Bargmann form on (M, τ, h), and $f \in \Omega^2(G(M))$ the mass torsion of ∇ with respect to a.*

(i) The mass torsion is explicitly given by

$$f = \mathrm{d}a + \delta_{ab}\varpi^a \wedge \theta^b, \tag{5.10}$$

where ϖ^a is the boost part of the connection form and $\theta = (\theta^A) \in \Omega^1_{\mathsf{Gal}}(G(M), \mathbb{R}^{n+1})$ is the tensorial form corresponding to the canonical solder form.

(ii) Let $\sigma = (v, \mathsf{e}_a)$ be a local Galilei frame, and consider the local representative $f = \sigma^* f$ of the mass torsion.
Under local rotations of Galilei frames $\sigma \mapsto (v, \mathsf{e}_b(R^{-1})^b{}_a)$, the local representative f is invariant; under local Galilei boosts $\sigma = (v, \mathsf{e}_a) \mapsto (v - \mathsf{e}_a k^a, \mathsf{e}_a)$, it transforms according to

$$f \mapsto f + k_a T^a + \frac{1}{2}|k|^2 \mathrm{d}\tau. \tag{5.11}$$

Here $T = v \otimes T^t + \mathsf{e}_a \otimes T^a$ is the local frame form of the torsion of ∇, understood as a TM-valued two-form.

---

[4] This name derives from the central direction of the Bargmann algebra being related to the mass of massive 'Galilei-invariant' (in fact locally Gal $\times$ U(1)-invariant, as we will see later) matter fields.
[5] In my (the author's) article [12], I also used them for the local representatives instead.

Proof (i) Using the explicit form (5.3) of $\dot{\rho}'$, the extended torsion is given by

$$
\begin{aligned}
(T^A, f) &= \mathrm{d}^\omega(\theta^A, a) \\
&= \mathrm{d}(\theta^A, a) + \dot{\rho}'_{(\omega, \varpi)} \wedge (\theta^A, a) \\
&= (\mathrm{d}\theta^A + \omega^A{}_B \wedge \theta^B, \mathrm{d}a + \delta_{ab}\varpi^a \wedge \theta^b).
\end{aligned}
\tag{5.12}
$$

(ii) This follows from the extended torsion being $\dot{\rho}$-tensorial, in direct parallel to the derivation of the transformation behaviour of a in Construction 5.5. $\qquad\square$

Remark 5.8 Consider a Galilei manifold (M, τ, h) with a Bargmann form a and a Galilei connection. As we have seen in Construction 5.5 and Proposition 5.7 (ii), the local representatives $a = \sigma^* a$ of the Bargmann form and $f = \sigma^* f$ of the mass torsion with respect to a Galilei frame $\sigma = (v, e_a)$ are invariant under rotations of the spatial frame fields e_a. This means that the local representatives only depend on the choice of unit timelike vector field v. Therefore, we will from now on also speak of the local representatives of a Bargmann form and/or the mass torsion with respect to a unit timelike vector field.

Under a Milne boost $v \mapsto v - w$ (with spacelike vector field w), these local representatives transform according to

$$
a \mapsto a + \underset{v}{h}(w, \cdot) + \frac{1}{2}|w|^2 \tau,
\tag{5.13a}
$$

$$
f \mapsto f + \underset{v}{h}(w, T(\cdot, \cdot)) + \frac{1}{2}|w|^2 \mathrm{d}\tau,
\tag{5.13b}
$$

where $|w|^2 = {}^{(n)}h(w, w)$. (These are (5.8), (5.11) expressed in terms of the Milne boost field $w = k^a e_a$.) $\qquad\triangle$

Corollary 5.9 *On a Galilei manifold with Bargmann form, the Newton–Coriolis form of a Galilei connection with respect to a unit timelike vector field v may be expressed in terms of the local representatives of the Bargmann form and the mass torsion with respect to v as*

$$
\Omega = f - \mathrm{d}a.
\tag{5.14}
$$

Proof Taking the pullback of Proposition 5.7 (i) along a local Galilei frame $\sigma = (v, e_a)$ whose timelike component is the given v, we obtain $f = \mathrm{d}a + \delta_{ab}\varpi^a \wedge e^b$. Combining this with Proposition 4.14 (which stated $\Omega = \delta_{ab}\varpi^a \wedge e^b$), the result follows. $\qquad\square$

Theorem 5.10 *On a Galilei manifold with a chosen Bargmann form, Galilei connections are uniquely determined by their extended torsion.*

In particular, if the Galilei manifold has absolute time, there is a unique Galilei connection with vanishing extended torsion. This connection is Newtonian.

Proof Since according to the classification theorem (Theorem 2.29) Galilei connections are determined by their torsion and Newton–Coriolis form, this follows directly from Corollary 5.9.

In the case of vanishing extended torsion, the connection being Newtonian follows by $d\Omega = -dda = 0$ (using Theorem 2.43). $\qquad\qquad\qquad\qquad\qquad\qquad\square$

Thus the notion of Bargmann form has allowed us to formulate a *global* version of the classification theorem for Galilei connections, with the extended torsion combining the torsion and the (local) Newton–Coriolis forms into a global object.

Construction 5.11 We may also reconstruct a Bargmann form from its local representatives with respect to unit timelike vector fields, as follows.

Let (M, τ, h) be a Galilei manifold, and consider an open cover $\{U_i\}_{i\in I}$ of M together with a choice of unit timelike vector field v_i and one-form a_i on each open set U_i, in such a way that the forms satisfy the transformation behaviour (5.13a) on the overlaps. Then the forms 'patch up' to a unique Bargmann form $\boldsymbol{a}$ whose local representative with respect to each v_i is a_i: we may cover each U_i by open sets on which Galilei frames are defined whose timelike vector field is v_i. Thus, we obtain an open cover of M by domains of definitions of Galilei frames; and for each of these frames, we have a one-form we want to be the local representative of the to-be-defined Bargmann form. We need these forms to satisfy the transformation behaviour of local representatives of a Bargmann form under the respective local Galilei transformations. However, for the frames corresponding to the same v_i, the forms coincide; and under frame change between frames corresponding to different vector fields, the forms satisfy the correct transformation behaviour by assumption. Hence, we indeed obtain a Bargmann form as desired. $\qquad\qquad\triangle$

Remark 5.12 Consider a Newtonian manifold (M, τ, h, ∇) on which we have a globally defined unit timelike vector field v such that the Newton–Coriolis form of ∇ with respect to v is exact, i.e. we can write $\Omega = -da$. By Construction 5.11 we may lift this a to a Bargmann form $\boldsymbol{a}$. The corresponding extended-torsion-free Galilei connection is then the original Newtonian connection ∇: considering the mass torsion of ∇ with respect to $\boldsymbol{a}$, its local representative with respect to v is $f = da + \Omega = da - da = 0$, so ∇ has vanishing extended torsion.

If Ω is not exact, then by the Poincaré lemma we still get local forms a such that $\Omega = -da$ (since Ω is closed by ∇ being Newtonian), but those don't 'patch up' to a global Bargmann form.

Note that when obtaining a Newtonian manifold as a formal $c \to \infty$ limit of a Lorentzian manifold as in Theorem 3.50, according to Remark 3.51 the Newton–Coriolis form of the limiting connection ∇ with respect to the unit timelike vector field $\hat{v}$ that naturally arises from the formal power series expansion in c^{-1} is of the form $\hat{\Omega} = \tau \wedge d\hat{\phi} = -d(\hat{\phi}\tau)$. So in this case we obtain a Bargmann form with local representative $\hat{a} = \hat{\phi}\tau$.

The same is true for Newtonian spacetimes with absolute rotation that arise from geometrisation of solutions of standard Newtonian gravity, i.e. by performing the converse construction of the Trautman recovery theorem (Theorem 3.40) as in Construction 3.32: the constructed Newton–Coriolis form is again $\Omega = \tau \wedge d\phi = -d(\phi\tau)$, so we get a Bargmann form with $a = \phi\tau$. $\triangle$

In particular, combined with Theorem 5.10 this shows the following:

Theorem 5.13 *Let (M, τ, h) be a Galilei manifold and ∇ a Galilei connection on it. Then ∇ is Newtonian if and only if for any point $p \in M$, there exists a Bargmann form $\boldsymbol{a}$ on an open neighbourhood of p such that the extended torsion of ∇ with respect to $\boldsymbol{a}$ vanishes.*

Put differently: the Newtonian connections are precisely those Galilei connections which are locally extended-torsion-free. $\square$

Note that this is somewhat similar to the characterisation of Newtonian connections as the locally special torsion-free Galilei connections from Theorem 2.44.

Construction 5.14 Let (M, τ, h) be a Galilei manifold, $G(M)$ its Galilei frame bundle, and $\varphi \colon M \to N$ a diffeomorphism. By pushforward, we obtain a Galilei manifold $(N, \varphi_*\tau, \varphi_*h)$, whose Galilei frame bundle we denote by $G(N)$. By acting on Galilei bases of (M, τ, h) with the differential of φ, we can lift φ to a principal bundle isomorphism $\boldsymbol{\varphi} \colon G(M) \overset{\cong}{\to} G(N)$:

$$G_p(M) \ni (e_A) \overset{\varphi}{\mapsto} (D\varphi|_p(e_A)) \in G_{\varphi(p)}(N) \tag{5.15}$$

Put differently, $\boldsymbol{\varphi}$ is the natural lift of φ to an isomorphism of the linear frame bundles $F(M)$ and $F(N)$, restricted to the Galilei frame bundles.

Let now $\theta \in \Omega^1_{\mathrm{Gal}}(G(M), \mathbb{R}^{n+1})$ be the tensorial form corresponding to the canonical solder form of $G(M) \times_{\mathrm{Gal}} \mathbb{R}^{n+1}$. By construction, the pushforward $\varphi_*\theta \in \Omega^1_{\mathrm{Gal}}(G(N), \mathbb{R}^{n+1})$ is again tensorial, and corresponds to the canonical solder form of $G(N) \times_{\mathrm{Gal}} \mathbb{R}^{n+1}$. Therefore, given a Bargmann form $\boldsymbol{a}$ on (M, τ, h), its pushforward $\varphi_*\boldsymbol{a} \in \Omega^1(G(N))$ along $\boldsymbol{\varphi}$ is a Bargmann form on $(N, \varphi_*\tau, \varphi_*h)$: the condition of $(\theta, \boldsymbol{a})$ being tensorial translates directly to $(\varphi_*\theta, \varphi_*\boldsymbol{a})$ being tensorial.

Locally, this is represented by pushforward along φ: given a unit timelike vector field v on (M, τ, h), the pushforward φ_*v is a unit timelike vector field on $(N, \varphi_*\tau, \varphi_*h)$; and if a is the local representative of $\boldsymbol{a}$ with respect to v, then the local representative of $\varphi_*\boldsymbol{a}$ with respect to φ_*v is φ_*a. $\triangle$

Construction 5.15 Let (M, τ, h) be a Galilei manifold and $\boldsymbol{a} \in \Omega^1(G(M))$ a Bargmann form on it.[6] Given a unit timelike vector field v on (M, τ, h), let a be the local representative of $\boldsymbol{a}$ with respect to v, and consider the unit timelike vector field

$$\hat{v} := v + h(a, \cdot), \tag{5.16a}$$

i.e. in components

$$\hat{v}^\mu := v^\mu + h^{\mu\nu} a_\nu . \tag{5.16b}$$

Under a Milne boost $v \mapsto v - w$ for some spacelike vector field w, the local representative of the Bargmann form transforms according to (5.13a). Thus, under this change of v the vector field $\hat{v}$ transforms as

$$\begin{aligned}
\hat{v}^\mu \mapsto\ & v^\mu - w^\mu + h^{\mu\nu}\left(a_\nu + h_{\nu\rho}w^\rho + \frac{1}{2}|w|^2\tau_\nu\right) \\
=\ & v^\mu - w^\mu + h^{\mu\nu}a_\nu + P^\mu_\rho w^\rho \\
=\ & v^\mu + h^{\mu\nu}a_\nu \\
=\ & \hat{v}^\mu .
\end{aligned} \tag{5.17}$$

This shows that $\hat{v}$ is *boost-invariant* in the following sense: replacing v by any other unit timelike vector field, i.e. applying a Milne boost to it, the field $\hat{v}$ defined by (5.16) will be the same. Put differently, this means that $\hat{v}$ is determined just by τ, h, and $\boldsymbol{a}$.

Since $\hat{v}$ is itself a unit timelike vector field, it makes sense to speak of the representative of the Bargmann form $\boldsymbol{a}$ with respect to $\hat{v}$. This representative we denote by $\hat{a}$. Note that, by definition, this $\hat{a}$ is—just as $\hat{v}$—fully determined by τ, h, and $\boldsymbol{a}$.

We may compute $\hat{a}$ as follows: for any (fixed, but arbitrary) unit timelike v with corresponding representative a of $\boldsymbol{a}$, we may view the boost-invariant $\hat{v}$ as 'just another unit timelike vector field' that arises from v by applying a specific Milne boost, namely as $\hat{v} = v - \hat{w}$ with $\hat{w} = -h(a, \cdot)$. Hence, according to (5.13a) $\hat{a}$ is given by

$$\begin{aligned}
\hat{a}_\mu &= a_\mu + h_{\mu\nu}\hat{w}^\nu + \frac{1}{2}|\hat{w}|^2\tau_\mu \\
&= a_\mu - h_{\mu\nu}h^{\nu\rho}a_\rho + \frac{1}{2}h(a, a)\tau_\mu \\
&= a_\mu - P^\rho_\mu a_\rho + \frac{1}{2}h(a, a)\tau_\mu \\
&= v^\rho\tau_\mu a_\rho + \frac{1}{2}h(a, a)\tau_\mu .
\end{aligned} \tag{5.18}$$

This shows that

$$\hat{a} = \hat{\phi}\tau \tag{5.19a}$$

---

[6] Some of the discussion in this construction is in parallel to parts of my (the author's) article [2].

with

$$\hat{\phi} = a(v) + \frac{1}{2}h(a, a). \tag{5.19b}$$

Since we know that $\hat{a}$ is determined by τ, h, and $\boldsymbol{a}$, we know that the function $\hat{\phi}$ appearing in (5.19) is determined by these fields as well. Thus $\hat{\phi}$ is boost-invariant in the same sense as $\hat{v}$: the explicit formula (5.19b) yields the same result for any choice of v.

Note that the representative $\hat{a}$ of $\boldsymbol{a}$ with respect to the boost-invariant unit timelike vector field $\hat{v}$ is proportional to τ. Conversely, this may be used to *define* $\hat{v}$: for τ, h, $\boldsymbol{a}$ given, $\hat{v}$ is the *unique* unit timelike vector field for which the local representative of $\boldsymbol{a}$ is proportional to τ (Exercise 5.16).

Similarly to the computation of $\hat{a}$, we can compute the covariant space metric $h_{\hat{v}}$ with respect to $\hat{v}$, whose components we denote by $\hat{h}_{\mu\nu}$: according to (2.52b), it is given by

$$\begin{aligned}
\hat{h}_{\mu\nu} &= h_{\mu\nu} + \hat{w}_\mu \tau_\nu + \tau_\mu \hat{w}_\nu + |\hat{w}|^2 \tau_\mu \tau_\nu \\
&= h_{\mu\nu} - P^\rho_\mu a_\rho \tau_\nu - \tau_\mu P^\sigma_\nu a_\sigma + h(a, a)\tau_\mu \tau_\nu \\
&= h_{\mu\nu} - a_\mu \tau_\nu - \tau_\mu a_\nu + (2a(v) + h(a, a))\tau_\mu \tau_\nu \\
&= h_{\mu\nu} - 2a_{(\mu}\tau_{\nu)} + 2\hat{\phi}\tau_\mu \tau_\nu \ .
\end{aligned} \tag{5.20}$$

$\triangle$

Exercise 5.16 (*Characterisation of the boost-invariant vector field*) Let (M, τ, h) be a Galilei manifold and $\boldsymbol{a} \in \Omega^1(G(M))$ a Bargmann form on it. Show that there is a unique unit timelike vector field for which the local representative of $\boldsymbol{a}$ is proportional to τ.

Hint: Start with an arbitrary unit timelike vector field v and corresponding representative a of $\boldsymbol{a}$. The representative $\tilde{a}$ of $\boldsymbol{a}$ with respect to any other unit timelike vector field $\tilde{v} = v - w$ (with w spacelike) is given by (5.13a). Show that the condition that this $\tilde{a}$ be proportional to τ uniquely determines w. $\triangle$

Remark 5.17 Note that Construction 5.15 shows that a Bargmann form $\boldsymbol{a} \in \Omega^1(G(M))$ on (M, τ, h) can equivalently be described by the pair $(\hat{v}, \hat{\phi})$. Since $\hat{v}$ and $\hat{\phi}$ are defined on the spacetime manifold M, it might seem easier to work with them instead of working with $\boldsymbol{a}$. However, in many applications it is actually more suitable to work in terms of the Bargmann form's local representative a with respect to an *arbitrary* unit timelike vector field v, which may be chosen appropriate to the situation at hand, or in terms of the Bargmann form $\boldsymbol{a}$ as a single global object on the Galilei frame bundle. $\triangle$

5.3   Bargmann Forms and Extension of Galilei Connections

In this section, we discuss an interpretation of Bargmann forms in terms of connections on an extension of the Galilei frame bundle. This is a 'globalised' formulation of the formalism discussed in the article [7].[7]

Construction 5.18 Let (M, τ, h) be a Galilei manifold, and $G(M)$ its Galilei frame bundle. We can extend $G(M)$ to a principal Barg-bundle as follows.

Denote by $\rho \colon \mathsf{Gal} \to \mathsf{Aut}(\mathbb{R}^{n+1} \times \mathbb{R})$ the group homomorphism from the definition of the Bargmann group (Definition 5.1), and by $\dot{\rho} \colon \mathsf{Gal} \to \mathsf{Aut}(\mathbb{R}^{n+1} \oplus \mathbb{R})$ the induced representation (5.2). Let further $\tilde{\rho} \colon \mathsf{Gal} \to \mathsf{Diff}(\mathsf{Barg})$ be the left action of Gal on Barg by multiplication (understanding Gal as a subgroup of Barg via $\mathsf{Gal} \ni (R, k) \mapsto (R, k, 0, 0) \in \mathsf{Barg}$). The associated bundle

$$B(M) := G(M) \times_{\tilde{\rho}} \mathsf{Barg} \tag{5.21}$$

is then a principal Barg-bundle (with action given by right multiplication). It extends $G(M)$ along the natural inclusion map

$$\gamma \colon G(M) \to B(M), \quad \gamma(p) = [p, \mathsf{e}_{\mathsf{Barg}}], \tag{5.22}$$

where $\mathsf{e}_{\mathsf{Barg}} = (\mathbb{1}, 0, 0, 0)$ is the neutral element of Barg.

In this situation, connections $\hat{\omega}$ on $B(M)$ are in one-to-one correspondence with pairs (ω, Θ) of connections ω and $\dot{\rho}$-tensorial one-forms $\Theta \in \Omega^1_{\dot{\rho}}(G(M), \mathbb{R}^{n+1} \oplus \mathbb{R})$ on $G(M)$ via the pullback condition

$$\gamma^* \hat{\omega} = (\omega, \Theta). \tag{5.23a}$$

Furthermore, the pullback of the curvature form $\hat{R}$ of $\hat{\omega}$ is given by the curvature form R of ω and the exterior covariant derivative of Θ with respect to ω,

$$\gamma^* \hat{R} = (R, \mathrm{d}^\omega \Theta). \tag{5.23b}$$

This generalises the classical situation of connections on the affine frame bundle of a manifold, and is generally true for 'semidirect extensions' of principal bundles as encountered here. A fully detailed discussion of this extension construction may be found in Appendix B.

Using this construction, we obtain a 'gauge-theoretic' interpretation of Bargmann forms: by definition, a Bargmann form a on (M, τ, h) together with the tensorial form $\theta \in \Omega^1_{\mathsf{Gal}}(G(M), \mathbb{R}^{n+1})$ corresponding to the canonical solder form combines into a $\dot{\rho}$-tensorial form $\Theta = (\theta, a)$. Together with a principal connection $\omega \in \Omega^1(G(M), \mathfrak{gal})$, i.e. with a Galilei connection, this Θ then gives a 'Bargmann

---

[7] Some of the discussion in this section is in great parallel to that in my (the author's) article [12].

connection' $\hat{\omega}$ on the extended bundle $B(M)$ via (5.23a). Note however that the notion of Bargmann form also makes sense without a choice of Galilei connection.

The extended torsion also acquires a 'gauge-theoretic' interpretation in this setting of the 'Bargmann extension' of the Galilei frame bundle: according to (5.23b), given a Galilei connection ∇ on (M, τ, h) with corresponding principal connection ω, its extended torsion with respect to a Bargmann form a corresponds to the $(\mathbb{R}^{n+1} \oplus \mathbb{R})$-valued part of the curvature of the 'Bargmann connection' $\hat{\omega}$ on $B(M)$ given by ω and $\Theta = (\theta, a)$. $\triangle$

In this context of the interpretation of Bargmann forms in terms of connections on the 'Bargmann extension' of the Galilei frame bundle, we obtain a notion of $\mathbb{R}$ gauge transformations for Bargmann forms in a natural way, corresponding to the central $\mathbb{R}$ direction in the Bargmann group:

Construction 5.19 Let (M, τ, h) be a Galilei manifold, $G(M) \xrightarrow{\pi} M$ its Galilei frame bundle, and $B(M) \xrightarrow{\hat{\pi}} M$ the principal Barg-bundle extending $G(M)$ as discussed in Construction 5.18. Given any smooth function $\chi \in C^\infty(M)$, we obtain a gauge transformation f_χ of $B(M)$, i.e. a fibre-preserving principal bundle automorphism $f_\chi : B(M) \to B(M)$, as

$$f_\chi(p) := p \cdot (\mathbb{1}, 0, 0, \chi \circ \hat{\pi}). \tag{5.24a}$$

Given a connection $\hat{\omega}$ on $B(M)$, we may act on it (actively!) with the gauge transformation f_χ, giving the new connection

$$f_\chi^* \hat{\omega} = \hat{\omega} + (0, 0, 0, \hat{\pi}^*(\mathrm{d}\chi)). \tag{5.24b}$$

For the case of $\hat{\omega}$ given by a connection ω on $G(M)$ and a Bargmann form a on (M, τ, h), this means that applying the gauge transformation given by χ, we obtain a new Bargmann form

$$a \mapsto a + \pi^*(\mathrm{d}\chi), \tag{5.25a}$$

which on M is locally represented by

$$a \mapsto a + \mathrm{d}\chi. \tag{5.25b}$$

These transformations (5.25) are called *type I gauge transformations* of Bargmann forms.

Note that with respect to two Bargmann forms which are related to each other by such a type I gauge transformation, a Galilei connection has the same extended torsion, since the local representative of the mass torsion satisfies $f = \mathrm{d}a + \Omega = \mathrm{d}(a + \mathrm{d}\chi) + \Omega$.

In particular, the Bargmann forms constructed in Remark 5.12 from Newtonian connections with globally exact Newton–Coriolis forms are unique up to type I gauge transformations. $\triangle$

Remark 5.20 The interpretation of Bargmann forms in terms of connections on a Barg extension of the Galilei frame bundle is very natural—indeed it is the geometric context in which Bargmann forms were historically first discovered [4–6].

In contemporary literature that uses terminology inspired from quantum field theory, this procedure of considering a connection valued in the Bargmann algebra— and thus introducing a kind of local Bargmann symmetry—is commonly called 'gauging the Bargmann algebra' [1,7]. Therefore, this literature sometimes uses the phrasing that Galilei geometry 'can be obtained from gauging the Bargmann algebra'.

However, despite the naturalness of this point of view, as we will see later it is not the natural one when considering the Newtonian limit from Lorentzian to Galilei geometry, at least when allowing $d\tau \neq 0$. $\triangle$

5.4   Variational Description of Coupling to Massive Matter

In this section, we will show how Bargmann forms allow to formulate variational principles for the dynamical equations of massive matter coupled to Newton–Cartan gravity.

Construction 5.21 (*The massive point particle*) Let (M, τ, h) be a Galilei manifold with a Bargmann form $\boldsymbol{a}$. We consider the following action functional for a timelike future-directed worldline γ:

$$
\begin{aligned}
S[\gamma] &= \int_{\lambda_i}^{\lambda_f} d\lambda \left(\frac{1}{2} m \frac{h(\gamma'(\lambda), \gamma'(\lambda))}{\tau(\gamma'(\lambda))} - m a(\gamma'(\lambda)) \right) \\
&= \int_{\gamma} \left(\frac{1}{2} m \frac{h(\cdot, \cdot)}{\tau} - m a \right)
\end{aligned}
\tag{5.26}
$$

Here $m \in \mathbb{R}$ is a parameter interpreted as a mass, v is a unit timelike vector field, and a is the local representative of $\boldsymbol{a}$ with respect to v. Note that the first summand in the integrand in fact defines a one-form along the curve, since for any $\mu \in \mathbb{R}$ we have $\frac{h(\mu \cdot \gamma', \mu \cdot \gamma')}{\tau(\mu \cdot \gamma')} = \mu \cdot \frac{h(\gamma', \gamma')}{\tau(\gamma')}$. Since the action is an integral over a one-form, it is reparametrisation invariant, i.e. depends on γ only through its image.

The expression (5.26) for the action explicitly depends on the choice of unit timelike vector field v: both the covariant space metric h and the local representative a of the Bargmann form depend on it. In fact, however, the action is independent of the choice of frame: according to Proposition 2.37 and Remark 5.8, under a Milne boost $v \mapsto v - w$, h and a transform as

$$
h \rightarrow h + h(w, \cdot) \otimes \tau + \tau \otimes h(w, \cdot) + |w|^2 \tau \otimes \tau,
\tag{5.27a}
$$

$$
a \rightarrow a + h(w, \cdot) + \frac{1}{2} |w|^2 \tau;
\tag{5.27b}
$$

therefore, the integrand transforms as

$$\frac{1}{2}m\frac{\underset{v}{h}(\gamma',\gamma')}{\tau(\gamma')} - ma(\gamma') \mapsto \frac{1}{2}m\left(\frac{\underset{v}{h}(\gamma',\gamma')}{\tau(\gamma')} + 2\underset{v}{h}(w,\gamma') + |w|^2\tau(\gamma')\right)$$

$$- m\left(a(\gamma') + \underset{v}{h}(w,\gamma') + \frac{1}{2}|w|^2\tau(\gamma')\right)$$

$$= \frac{1}{2}m\frac{\underset{v}{h}(\gamma',\gamma')}{\tau(\gamma')} - ma(\gamma'). \tag{5.27c}$$

In the case of a Newtonian manifold with absolute rotation that arose from a solution of the standard formulation of Newtonian gravity, with $a = \phi\tau$ (as in Remark 5.12), the action is just the usual action of a point particle in Newtonian mechanics, namely the integral over kinetic minus gravitational potential energy: parametrising γ by absolute time, i.e. as $\gamma(t)$ such that $\tau(\dot\gamma(t)) = 1$, we have $S = \int \mathrm{d}t(\frac{1}{2}m\delta_{ab}\dot\gamma^a\dot\gamma^b - m\phi)$. Thus one might say that the action (5.26) arises from the action of a particle in Newtonian mechanics by 'geometrisation'.

Note that the action (5.26) is not invariant under type I gauge transformations of the Bargmann form (Construction 5.19). However, under a type I gauge transformation by $\chi \in C^\infty(M)$, it transforms as $S[\gamma] \mapsto S[\gamma] - m\int_\gamma \mathrm{d}\chi = S[\gamma] - m$ $(\chi(\gamma_\mathrm{f}) - \chi(\gamma_\mathrm{i}))$, i.e. by a boundary term (such that the Euler–Lagrange equations of the action stay the same).

Now we want to determine the equations of motion/Euler–Lagrange equations that arise from the above action by demanding it to be stationary under variations of γ with fixed endpoints. Computing the variation of the action, we first obtain

$$\delta S = \int \mathrm{d}\lambda\left[\frac{1}{2}m\frac{(\delta h_{\mu\nu})\gamma'^\mu\gamma'^\nu}{\tau(\gamma')} + m\frac{h_{\mu\nu}\gamma'^\mu\delta\gamma'^\nu}{\tau(\gamma')} - m(\delta a_\mu)\gamma'^\mu - ma_\mu\delta\gamma'^\mu\right.$$

$$\left. - \frac{1}{2}m\frac{\underset{v}{h}(\gamma',\gamma')}{(\tau(\gamma')^2)}\left((\delta\tau_\sigma)\gamma'^\sigma + \tau_\nu\delta\gamma'^\rho\right)\right]$$

$$= \int \mathrm{d}\lambda\left[\frac{1}{2}m\delta\gamma^\rho\frac{(\partial_\rho h_{\mu\nu})\gamma'^\mu\gamma'^\nu}{\tau(\gamma')} - m\delta\gamma^\nu\frac{\mathrm{d}}{\mathrm{d}\lambda}\left(\frac{h_{\mu\nu}\gamma'^\mu}{\tau(\gamma')}\right)\right.$$

$$- m\delta\gamma^\rho(\partial_\rho a_\mu)\gamma'^\mu + m\gamma'^\rho(\partial_\rho a_\mu)\delta\gamma^\mu - \frac{1}{2}m\frac{\underset{v}{h}(\gamma',\gamma')}{(\tau(\gamma')^2)}\delta\gamma^\rho(\partial_\rho\tau_\sigma)\gamma'^\sigma$$

$$\left. + \frac{1}{2}m\delta\gamma^\rho\frac{\mathrm{d}}{\mathrm{d}\lambda}\left(\frac{\underset{v}{h}(\gamma',\gamma')}{(\tau(\gamma')^2)}\tau_\rho\right)\right]$$

$$= \int \mathrm{d}\lambda\left[\frac{1}{2}m\delta\gamma^\rho\frac{(\partial_\rho h_{\mu\nu})\gamma'^\mu\gamma'^\nu}{\tau(\gamma')} - m\delta\gamma^\nu\frac{\mathrm{d}}{\mathrm{d}\lambda}\left(\frac{h_{\mu\nu}\gamma'^\mu}{\tau(\gamma')}\right) + m\delta\gamma^\rho(\mathrm{d}a)_{\mu\rho}\gamma'^\mu\right.$$

$$\left. - \frac{1}{2}m\frac{\underset{v}{h}(\gamma',\gamma')}{(\tau(\gamma')^2)}\delta\gamma^\rho(\partial_\rho\tau_\sigma)\gamma'^\sigma + \frac{1}{2}m\delta\gamma^\rho\frac{\mathrm{d}}{\mathrm{d}\lambda}\left(\frac{\underset{v}{h}(\gamma',\gamma')}{(\tau(\gamma')^2)}\tau_\rho\right)\right], \tag{5.28}$$

where we have used partial integration and vanishing of the variation $\delta\gamma^\mu$ on the boundary to express all terms involving the derivative $\delta\gamma'^\mu$ in terms of $\delta\gamma^\mu$ itself. Continuing the calculation, we have

$$\frac{\mathrm{d}}{\mathrm{d}\lambda}\left(\frac{h_{\mu\nu}\gamma'^{\mu}}{\tau(\gamma')}\right) = \frac{1}{\tau(\gamma')}(\partial_\rho h_{\mu\nu})\gamma'^\rho\gamma'^\mu$$

$$+ \frac{1}{\tau(\gamma')}h_{\mu\nu}\gamma''^{\mu} + h_{\mu\nu}\gamma'^{\mu}\frac{\mathrm{d}}{\mathrm{d}\lambda}\left(\frac{1}{\tau(\gamma')}\right), \tag{5.29a}$$

$$\frac{\mathrm{d}}{\mathrm{d}\lambda}\left(\frac{\underset{v}{h}(\gamma',\gamma')}{(\tau(\gamma')^2)}\tau_\rho\right) = \frac{\underset{v}{h}(\gamma',\gamma')}{(\tau(\gamma')^2)}(\partial_\sigma\tau_\rho)\gamma'^\sigma + \tau_\rho\frac{\mathrm{d}}{\mathrm{d}\lambda}\left(\frac{\underset{v}{h}(\gamma',\gamma')}{(\tau(\gamma')^2)}\right). \tag{5.29b}$$

For the respective last terms of these expressions we further obtain

$$\frac{\mathrm{d}}{\mathrm{d}\lambda}\left(\frac{1}{\tau(\gamma')}\right) = -\frac{1}{(\tau(\gamma'))^2}\frac{\mathrm{d}\tau(\gamma')}{\mathrm{d}\lambda}, \tag{5.29c}$$

$$\frac{\mathrm{d}}{\mathrm{d}\lambda}\left(\frac{\underset{v}{h}(\gamma',\gamma')}{(\tau(\gamma')^2)}\right) = \frac{1}{(\tau(\gamma'))^2}\left((\partial_\kappa h_{\mu\nu})\gamma'^\kappa\gamma'^\mu\gamma'^\nu + 2h_{\mu\nu}\gamma'^\mu\gamma''^\nu\right)$$

$$- 2\frac{1}{(\tau(\gamma'))^3}\frac{\mathrm{d}\tau(\gamma')}{\mathrm{d}\lambda}\underset{v}{h}(\gamma',\gamma'). \tag{5.29d}$$

Into these we could further insert

$$\frac{\mathrm{d}\tau(\gamma')}{\mathrm{d}\lambda} = (\partial_\sigma\tau_\mu)\gamma'^\sigma\gamma'^\mu + \tau_\mu\gamma''^\mu, \tag{5.30}$$

but will refrain from doing so since the corresponding terms will not contribute to our end result.

Inserting (5.29) into (5.28), the action variation in the general case is given by

$$\delta S = \int \mathrm{d}\lambda\,\delta\gamma^\rho\left[-m\frac{1}{\tau(\gamma')}h_{\mu\rho}\gamma''^\mu - \frac{1}{2}m(2\partial_\nu h_{\mu\rho} - \partial_\rho h_{\mu\nu})\frac{\gamma'^\mu\gamma'^\nu}{\tau(\gamma')}\right.$$

$$+ m(\mathrm{d}a)_{\mu\rho}\gamma'^\mu - \frac{1}{2}m(\mathrm{d}\tau)_{\rho\sigma}\gamma'^\sigma\frac{\underset{v}{h}(\gamma',\gamma')}{(\tau(\gamma')^2)} + mh_{\mu\rho}\gamma'^\mu\frac{1}{(\tau(\gamma'))^2}\frac{\mathrm{d}\tau(\gamma')}{\mathrm{d}\lambda}$$

$$+ \frac{1}{2}m\tau_\rho\left(\frac{1}{(\tau(\gamma'))^2}\left((\partial_\kappa h_{\mu\nu})\gamma'^\kappa\gamma'^\mu\gamma'^\nu + 2h_{\mu\nu}\gamma'^\mu\gamma''^\nu\right)\right.$$

$$\left.\left. - 2\frac{1}{(\tau(\gamma'))^3}\frac{\mathrm{d}\tau(\gamma')}{\mathrm{d}\lambda}\underset{v}{h}(\gamma',\gamma')\right)\right]. \tag{5.31}$$

We now assume that γ is parametrised by time along γ as defined by τ, i.e. we have a parametrisation $\gamma(t)$ with $\tau(\dot{\gamma}(t)) = 1$. In particular this implies $\frac{\mathrm{d}\tau(\dot{\gamma})}{\mathrm{d}t} = 0$. Therefore, the equation of motion equivalent to $\delta S = 0$ that we may read off from (5.31) simplifies to

$$0 = h_{\mu\rho}\ddot{\gamma}^\mu + \frac{1}{2}(2\partial_\nu h_{\mu\rho} - \partial_\rho h_{\mu\nu})\dot{\gamma}^\mu\dot{\gamma}^\nu - (\mathrm{d}a)_{\mu\rho}\dot{\gamma}^\mu + \frac{1}{2}(\mathrm{d}\tau)_{\rho\sigma}\dot{\gamma}^\sigma\underset{v}{h}(\dot{\gamma},\dot{\gamma})$$

$$- \frac{1}{2}\tau_\rho\left((\partial_\kappa h_{\mu\nu})\dot{\gamma}^\kappa\dot{\gamma}^\mu\dot{\gamma}^\nu + 2h_{\mu\nu}\dot{\gamma}^\mu\ddot{\gamma}^\nu\right). \tag{5.32}$$

Raising the free index, we obtain

$$0 = P^\rho_\sigma \ddot{\gamma}^\sigma + \frac{1}{2} h^{\rho\sigma} (2\partial_\nu h_{\mu\sigma} - \partial_\sigma h_{\mu\nu}) \dot{\gamma}^\mu \dot{\gamma}^\nu - (\mathrm{d}a)_\mu{}^\rho \dot{\gamma}^\mu + \frac{1}{2} (\mathrm{d}\tau)^\rho{}_\sigma \dot{\gamma}^\sigma \underset{v}{h}(\dot{\gamma}, \dot{\gamma}).$$

$$(5.33)$$

Let now ∇ be an arbitrary Galilei connection on (M, τ, h). Using the classification theorem (Theorem 2.29), we obtain

$$P^\rho_\sigma \Gamma^\sigma_{\mu\nu} \dot{\gamma}^\mu \dot{\gamma}^\nu = \frac{1}{2} h^{\rho\sigma} (2\partial_\nu h_{\mu\sigma} - \partial_\sigma h_{\mu\nu}) \dot{\gamma}^\mu \dot{\gamma}^\nu - T_{\mu\nu}{}^\rho \dot{\gamma}^\mu \dot{\gamma}^\nu + \Omega_\mu{}^\rho \dot{\gamma}^\mu. \quad (5.34)$$

Comparing this to the equation of motion (5.33) and using $\Omega = f - \mathrm{d}a$ as well as $\ddot{\gamma}^\sigma + \Gamma^\sigma_{\mu\nu} \dot{\gamma}^\mu \dot{\gamma}^\nu = (\nabla_{\dot{\gamma}} \dot{\gamma})^\sigma$, we see that the equation of motion is equivalent to

$$0 = P^\rho_\sigma (\nabla_{\dot{\gamma}} \dot{\gamma})^\sigma + T_{\mu\nu}{}^\rho \dot{\gamma}^\mu \dot{\gamma}^\nu - f_\mu{}^\rho \dot{\gamma}^\mu + \frac{1}{2} (\mathrm{d}\tau)^\rho{}_\sigma \dot{\gamma}^\sigma \underset{v}{h}(\dot{\gamma}, \dot{\gamma}). \quad (5.35)$$

Combined with $\tau(\nabla_{\dot{\gamma}} \dot{\gamma}) = \nabla_{\dot{\gamma}}(\tau(\dot{\gamma})) = \nabla_{\dot{\gamma}}(1) = 0$, we thus see that for curves parametrised by time along them (as defined by τ), the equation of motion arising from the above action is

$$(\nabla_{\dot{\gamma}} \dot{\gamma})^\rho = -T_{\mu\nu}{}^\rho \dot{\gamma}^\mu \dot{\gamma}^\nu + f_\mu{}^\rho \dot{\gamma}^\mu - \frac{1}{2} (\mathrm{d}\tau)^\rho{}_\sigma \dot{\gamma}^\sigma \underset{v}{h}(\dot{\gamma}, \dot{\gamma}). \quad (5.36)$$

In the case of an extended-torsion-free connection, this reduces to the geodesic equation. We have thus obtained the geodesic equation—i.e. the equation of motion for a massive test particle in Newton–Cartan gravity—from a variational principle. Note that the use of the Bargmann form was crucial for this. Morally speaking, the Bargmann form allows us to speak about the Newtonian gravitational potential ϕ, but in proper geometric terms.

Note that even though by parametrising the curve by time along it we may write the action in the easier form

$$S[\gamma] = \int_{t_i}^{t_f} \mathrm{d}t \left(\frac{1}{2} m h_{\mu\nu} \dot{\gamma}^\mu \dot{\gamma}^\nu - m a_\mu \dot{\gamma}^\mu \right), \quad (5.37)$$

in the general case this is misleading for calculating the variation of the action, since for fixed endpoints of the curve the time along it may change under variation if $\mathrm{d}\tau \neq 0$. $\triangle$

Construction 5.22 (*The Schrödinger field*) The standard free 'non-relativistic' Schrödinger equation with mass m is

$$i\hbar\partial_t \Psi = -\frac{\hbar^2}{2m} \delta^{ab} \partial_a \partial_b \Psi \quad (5.38)$$

for a function $\Psi\colon \mathbb{R}^{n+1} \to \mathbb{C}$. It may be obtained by variation of the action functional

$$S_{\text{free}}[\Psi] = \int \mathrm{d}t \mathrm{d}^n x \left(\mathrm{i}\hbar m(\overline{\Psi}\, \partial_t \Psi - \overline{\partial_t \Psi}\, \Psi) - \hbar^2 \delta^{ab}\, \overline{\partial_a \Psi}\, \partial_b \Psi \right). \tag{5.39}$$

Here, we understand the Schrödinger equation as a *classical* field equation.[8]

We now want to couple the Schrödinger field to Newtonian gravity, described geometrically by Newton–Cartan gravity. In order to do so, we will 'geometrise' the above action, similar to how the point particle action in Construction 5.21 is a 'geometrised' version of the free point particle action. As a first step, we observe that the term $\delta^{ab}\, \overline{\partial_a \Psi}\, \partial_b \Psi$ can be translated onto a general Galilei manifold (M, τ, h) simply as $h^{\mu\nu}\, \overline{\partial_\mu \Psi}\, \partial_\nu \Psi$. For the time derivative $\partial_t \Psi$ however we need to introduce a time direction—i.e. we need to replace this somehow by a term related to $v^\mu \partial_\mu \Psi$, where v is a unit timelike vector field. This would then introduce a dependence on the choice of v, i.e. the action would not be geometric in the sense of just depending on the Galilei manifold itself.

This apparent problem may however be solved with inspiration coming from the fact that the free Schrödinger equation is invariant under Galilei boosts only if those are implemented in a non-trivial manner, including a specific phase factor arising from the Bargmann group (details of this will be studied in Exercise 5.24). This was realised by Duval and Künzle in a paper from 1984 [6]. Their solution, adapted to our notation, works as follows.

We consider a Bargmann form $\boldsymbol{a}$ on our Galilei manifold (M, τ, h), and its local representative a with respect to a chosen unit timelike vector field v. The $\mathfrak{u}(1)$-valued (local) one-form $\mathrm{i}\frac{m}{\hbar} a$ on M then defines a connection on the trivial principal $\mathsf{U}(1)$-bundle $M \times \mathsf{U}(1)$. Note that this connection depends on the choice of v—a Milne boost will lead to a different local form a, and therefore to a different $\mathsf{U}(1)$ connection! The Schrödinger field is now a section in the associated vector bundle $E = (M \times \mathsf{U}(1)) \times_{\mathsf{U}(1)} \mathbb{C}$ with respect to the defining representation of $\mathsf{U}(1)$ on $\mathbb{C}$. We have a $\mathsf{U}(1)$-covariant derivative operator D on E, induced by our connection, acting on sections $\Psi \in \Gamma(E)$ as

$$\mathrm{D}_\mu \Psi = \left(\partial_\mu + \mathrm{i}\frac{m}{\hbar} a_\mu \right) \Psi. \tag{5.40}$$

We now use *this* covariant derivative operator to 'minimally couple' the free Schrödinger action to Newton–Cartan gravity, leading to the action

$$S[\Psi] = \int \mathrm{vol}\left(\mathrm{i}\hbar m v^\mu (\overline{\Psi}\, \mathrm{D}_\mu \Psi - \overline{\mathrm{D}_\mu \Psi}\, \Psi) - \hbar^2 h^{\mu\nu}\, \overline{\mathrm{D}_\mu \Psi}\, \mathrm{D}_\nu \Psi \right), \tag{5.41}$$

where $\mathrm{vol} = \tau \wedge \mathrm{e}^1 \wedge \cdots \wedge \mathrm{e}^n$ (for any choice of Galilei frame) is the natural volume form of (M, τ, h). Even though it is not obvious from the construction, this action

---

[8] Of course, one might quantise the classical Schrödinger field, which would yield the usual 'second-quantised' description of an arbitrary number of non-interacting quantum particles.

is in fact *invariant* under Milne boosts (see Exercise 5.23), thus solving the above problem.

By construction, the action is also invariant under type I gauge transformations of the Bargmann form (Construction 5.19) when we let those act on $M \times U(1)$, to which E is associated, as well, via the group homomorphism $\mathbb{R} \to U(1)$, $\chi \mapsto e^{i\chi}$.

The Schrödinger equation we obtain from variation of this action (with respect to $\overline{\Psi}$) is

$$i\hbar v^{\mu} D_{\mu} \Psi + \frac{i\hbar}{2} \frac{1}{f_{\text{vol}}} (\partial_{\mu}(f_{\text{vol}} v^{\mu})) \Psi = -\frac{\hbar^2}{2m} \frac{1}{f_{\text{vol}}} \left(\partial_{\mu} + i\frac{m}{\hbar} a_{\mu} \right) (f_{\text{vol}} h^{\mu\nu} D_{\nu} \Psi),$$

$$(5.42a)$$

where f_{vol} is the coordinate component of the volume form, $\text{vol} = f_{\text{vol}} \, dx^{n+1}$. For a Galilei manifold with absolute time, we can further rewrite this as

$$i\hbar v^{\mu} D_{\mu} \Psi + \frac{i\hbar}{2} (\nabla_{\mu} v^{\mu}) \Psi = -\frac{\hbar^2}{2m} h^{\mu\nu} D_{\mu} D_{\nu} \Psi, \tag{5.42b}$$

where ∇ is *any* torsion-free Galilei connection and we have extended the action of D to E-valued one-forms in the natural way, i.e.

$$D_{\mu} \eta_{\nu} = \left(\nabla_{\mu} + i\frac{m}{\hbar} a_{\mu} \right) \eta_{\nu}. \tag{5.43}$$

Note that (5.42b) is indeed independent of the choice of (torsion-free) Galilei connection: $\nabla_{\mu} v^{\mu}$ is (proportional to) the expansion of v, which depends only on v and h (see Proposition 3.11); and on the right-hand side, we project onto space and thus see only the spatial part $\overset{(n)}{\nabla}$ of the connection, which is the spatial Levi-Civita connection (Corollary 2.23).

In the case of a Newtonian manifold with absolute rotation that arose from a solution of the standard formulation of Newtonian gravity, with $a = \phi\tau$ (as in Remark 5.12), in adapted coordinates we have $v = \partial_t$ and $h^{\mu\nu} = \delta_a^{\mu} \delta_b^{\nu} \delta^{ab}$, giving the Schrödinger equation

$$i\hbar \partial_t \Psi = -\frac{\hbar^2}{2m} \delta^{ab} \partial_a \partial_b \Psi + m\phi \Psi \tag{5.44}$$

including a Newtonian gravitational potential term. △

Exercise 5.23 (*Local boost invariance of the Schrödinger action*) Show by explicit computation that the action (5.41) for the Schrödinger field on a Galilei spacetime is invariant under Milne boosts. △

Exercise 5.24 (*Bargmann symmetry of the free Schrödinger equation*) From the construction of the Bargmann form framework and the invariance of the action, we know that the Schrödinger equation (5.42b) on any Galilei manifold (M, τ, h) with absolute time and with Bargmann structure a is symmetric under all of the following transformations:

(i) combined type I gauge transformations of a and $\mathsf{U}(1)$ gauge transformations of Ψ,

(ii) Milne boosts of v and a,

(iii) pushforward of τ, h, v, a, Ψ by any diffeomorphism $\varphi\colon M \to M$.

In this exercise we will use this to show that the free Schrödinger equation on $\mathbb{R}^{n+1}$ (5.38) is symmetric under a specific action of the Bargmann group.

We consider the Newtonian manifold (M, τ, h, ∇) where $M = \mathbb{R}^{n+1}$ with affine coordinates (t, x^a), $\tau = \mathrm{d}t$, $h = \delta^{ab}\partial_a \otimes \partial_b$, and ∇ is the standard linear connection whose connection coefficients with respect to the coordinates (t, x^a) vanish.

(a) Show that any diffeomorphism $\varphi\colon \mathbb{R}^{n+1} \to \mathbb{R}^{n+1}$ under whose pushforward action τ, h, ∇ are invariant[9], i.e. $\varphi_*\tau = \tau$, $\varphi_*h = h$, $\varphi_*\nabla = \nabla$, is given by the standard action of an element (R, k, y) of the inhomogeneous Galilei group $\mathsf{Gal} \ltimes \mathbb{R}^{n+1}$ acting on $\mathbb{R}^{n+1}$, i.e.

$$\varphi(t, x^a) = (t + y^t, R^a{}_b x^b + t k^a + y^a). \tag{5.45}$$

Hint: From the condition $\varphi^*\nabla = \nabla$, you know that φ has to be an affine map of $\mathbb{R}^{n+1}$, i.e. that φ can be written as multiplication with a constant matrix followed by translation by a constant vector y. The remaining conditions then restrict the matrix to an element of Gal.

(b) We now consider the unit timelike vector field $v = \partial_t$. To this, we first apply a Milne boost parametrised by a spacelike vector field $\tilde{w}$ on (M, τ, h), and second pushforward by a Galilei transformation diffeomorphism $\varphi_{(R,k,y)}$ as in part (a). Show that for v to be invariant under this combined action, i.e. for

$$(\varphi_{(R,k,y)})_*(v - \tilde{w}) = v, \tag{5.46}$$

we need $\tilde{w} = (R^{-1})^a{}_b k^b \partial_a$.

(c) We consider the Bargmann form a on (M, τ, h) whose local representative with respect to $v = \partial_t$ vanishes, $a = 0$. As in part (b), to the local representative of our Bargmann form we apply a Milne boost by $(R^{-1})^a{}_b k^b \partial_a$, followed by pushforward by $\varphi_{(R,k,y)}$. This defines an action of the inhomogeneous Galilei group on Bargmann forms on our (M, τ, h).[10] Compute the local representative of the resulting new Bargmann form.

(d) Before the action of part (c), we now also apply a type I gauge transformation parametrised by $\chi \in C^\infty(M)$ to our original Bargmann form with $a = 0$. For which functions χ is the resulting Bargmann form $\tilde{a}$ equal to the original one, i.e. again locally given by $\tilde{a} = 0$?

---

[9] Note that invariance of ∇ will not be used in the rest of the exercise, it is just a nice way to obtain Galilei transformations.

[10] Since from parts (a) and (b) we know that the combined action (pushforward by $\varphi_{(R,k)}$) $\circ$ (Milne boost by $R^{-1}k$) leaves our τ, h, v invariant, this action on a really defines an action of the inhomogeneous Galilei group on Bargmann forms.

Hint: Evaluate the condition $0 = \tilde{a} = \varphi_*(\ldots)$ *in the equivalent form* $\cdots = 0$. *You should obtain* $\chi = -k_a R^a{}_b x^b + \frac{1}{2}|k|^2 t + \text{const}$.

(e) Combining all the previous parts, we see that the structure (τ, h, v, a) is invariant under the combined action (pushforward by $\varphi_{(R,k,y)}$) $\circ$ (Milne boost by $R^{-1}k$) $\circ$ (type I gauge transformation by χ as in part (d)). How does this combination act on a Schrödinger field, understood geometrically as in Construction 5.22? Show that this defines an action on Schrödinger fields by the Bargmann group. Use this to argue that the free Schrödinger equation (5.38) is symmetric under this action of the Bargmann group.

Hint: Write χ *in the form from the hint in part (d), and call the free constant* $-\beta$. *For the combination of two transformations, you have to perform a long computation in order to show you obtain the Bargmann group multiplication rule. The last part follows almost trivially from the previous considerations.* $\triangle$

5.5 Bargmann Forms from the Newtonian Limit of Lorentzian Geometry

In this section, we will explain how Bargmann forms may be included into the formal limiting process from Lorentzian geometry and GR to the geometry of Galilei manifolds and Newton–Cartan gravity, as presented in Sect. 3.4, by using a frame formulation.

Theorem 5.25 *Let (M, g) be a Lorentzian manifold, and let $(\mathrm{E}_A) = (\mathrm{E}_0, \mathrm{E}_a)$ be a local orthonormal frame with dual frame (E^A), such that the metric and inverse metric can be written as*

$$g = \eta_{AB}\mathrm{E}^A \otimes \mathrm{E}^B, \quad g^{-1} = \eta^{AB}\mathrm{E}_A \otimes \mathrm{E}_B, \tag{5.47}$$

where η_{AB} denotes the components of the Minkowski metric in Lorentzian coordinates, i.e. $(\eta_{AB}) = \mathrm{diag}(-1, 1, \ldots, 1)$. Assume that the frame and dual frame may be expanded as formal power series in c^{-1} as

$$\mathrm{E}^0 = c\tau + c^{-1}a + \mathrm{O}(c^{-2}), \qquad \mathrm{E}^a = \mathrm{e}^a + \mathrm{O}(c^{-1}), \tag{5.48a}$$

$$\mathrm{E}_0 = c^{-1}v + \mathrm{O}(c^{-2}), \qquad \mathrm{E}_a = \mathrm{e}_a + \mathrm{O}(c^{-2}) \tag{5.48b}$$

for some nowhere-vanishing one-form $\tau \in \Omega^1(M)$.[11]

---

[11] Note that for the formal expansions of E^0 and the E_a, we demand the leading order and next to leading order to differ by a factor of c^{-2}, while for E_0 and the E^a we only demand a factor of c^{-1}. If we were to demand a factor of c^{-2} also in the latter case, all following expansions in the statement of this theorem would proceed in steps of c^{-2} as well.

(i) *The expansions of the metric g and inverse metric g^{-1} satisfy the assumptions of Lemma 3.48, such that we obtain a Galilei manifold (M, τ, h) as a formal $c \to \infty$ limit of our Lorentzian manifold.*

(ii) *(v, e_a) is a local Galilei frame for the limiting Galilei manifold (M, τ, h), with dual frame (τ, e^a).*

(iii) *We consider a local Lorentz boost $(\Lambda^A{}_B)$ parametrised by the $\mathbb{R}^n$-valued boost velocity function k as*

$$\Lambda = \exp(\zeta^a K_a) \tag{5.49a}$$

with the rapidity

$$\zeta^a = \operatorname{artanh}(|k|/c)\frac{k^a}{|k|} \tag{5.49b}$$

and the boost generators

$$(K_a)^A{}_B = \delta_0^A \eta_{aB} - \delta_a^A \eta_{0B} \ . \tag{5.49c}$$

Explicitly, this means that

$$\Lambda^0{}_0 = 1 + c^{-2}\frac{|k|^2}{2} + O(c^{-4}), \tag{5.49d}$$

$$\Lambda^a{}_0 = c^{-1}k^a + O(c^{-3}) = \delta^{ab}\Lambda^0{}_b \ , \tag{5.49e}$$

$$\Lambda^a{}_b = \delta_b^a + c^{-2}\frac{k^a k_b}{2} + O(c^{-4}). \tag{5.49f}$$

Transforming the Lorentzian frame (E_A) by Λ according to

$$(\mathrm{E}_A) \mapsto (\tilde{\mathrm{E}}_A) = \left(\mathrm{E}_B(\Lambda^{-1})^B{}_A\right), \tag{5.50}$$

and expanding the new frame analogously to (5.48), we obtain a local Galilei boost of the Galilei frame (v, e_a) with boost velocity parameter k.
Furthermore, the local one-forms a that arise as the c^{-1} component of the timelike dual frame one-form E^0 transform as the local representatives of a Bargmann form on (M, τ, h) would, thereby in fact defining a Bargmann form $\boldsymbol{a}$.

(iv) *According to Lemma 3.48, the expansions of the metric and the inverse metric in c^{-1} naturally determine a unit timelike vector field $\hat{v} = -m(\tau, \cdot)$ on (M, τ, h), where $g^{-1} =: h + c^{-2}m + O(c^{-3})$, and a function $\hat{\phi}$ on M, by*

$$g = -c^2 \tau \otimes \tau + \underset{\hat{v}}{h} - 2\hat{\phi}\tau \otimes \tau + O(c^{-1}). \tag{5.51}$$

These have a direct interpretation in terms of the Bargmann form $\boldsymbol{a}$ obtained from the expansion of the frame in part (iii): $\hat{v}$ is the boost-invariant unit timelike *vector field determined by the Bargmann form $\boldsymbol{a}$ according to Construction 5.15, and $\hat{\phi}$ the corresponding function on M determined by $\hat{a} = \hat{\phi}\tau$ (where $\hat{a}$ is the local representative of $\boldsymbol{a}$ with respect to $\hat{v}$).[12]*

(v) Let $\overset{\text{L}}{\nabla}$ be a Lorentzian metric connection with a regular formal $c \to \infty$ limit (i.e. its coordinate components with respect to c-independent coordinates, or equivalently its local connection form with respect to the frame (v, e_a), have regular limits). Then its local connection form $(\overset{\text{L}}{\omega}{}^A{}_B)$ with respect to (E_A) expands as

$$\overset{\text{L}}{\omega}{}^0{}_0 = 0, \tag{5.52a}$$

$$\overset{\text{L}}{\omega}{}^a{}_0 = c^{-1}\varpi^a + \mathrm{O}(c^{-2}) = \delta^{ab}\overset{\text{L}}{\omega}{}^0{}_b, \tag{5.52b}$$

$$\overset{\text{L}}{\omega}{}^a{}_b = \omega^a{}_b + \mathrm{O}(c^{-1}), \tag{5.52c}$$

for local one-forms $\omega^a{}_b$, ϖ^a. Under local rotations and boosts of the frame, the $(\omega^a{}_b, \varpi^a)$ transform as the local connection form of a Galilei connection on (M, τ, h) would, thereby defining a Galilei connection ∇.

The torsion $\overset{\text{L}}{T}$ of $\overset{\text{L}}{\nabla}$ then expands as

$$\overset{\text{L}}{T}{}^0 = c\mathrm{d}\tau + c^{-1}f + \mathrm{O}(c^{-2}), \tag{5.53a}$$

$$\overset{\text{L}}{T}{}^a = T^a + \mathrm{O}(c^{-1}) \tag{5.53b}$$

in terms of the extended torsion (T, f) of ∇ with respect to the Bargmann form $\boldsymbol{a}$ obtained from the expansion of the frame in part (iii).[13] Here the components of $\overset{\text{L}}{T}$ are taken with respect to (E_A) and those of T with respect to (v, e_a), i.e. $\overset{\text{L}}{T} = \text{E}_0 \otimes \overset{\text{L}}{T}{}^0 + \text{E}_a \otimes \overset{\text{L}}{T}{}^a$ and $T = v \otimes \mathrm{d}\tau + \text{e}_a \otimes T^a$.

Similarly, in terms of the local curvature form $(R^A{}_B)$ of ∇ with respect to (e_A), the local curvature form $(\overset{\text{L}}{R}{}^A{}_B)$ of $\overset{\text{L}}{\nabla}$ with respect to (E_A) expands as

$$\overset{\text{L}}{R}{}^0{}_0 = \mathrm{O}(c^{-3}), \tag{5.54a}$$

$$\overset{\text{L}}{R}{}^a{}_0 = c^{-1}R^a{}_t + \mathrm{O}(c^{-2}) = \delta^{ab}\overset{\text{L}}{R}{}^0{}_b, \tag{5.54b}$$

$$\overset{\text{L}}{R}{}^a{}_b = R^a{}_b + \mathrm{O}(c^{-1}). \tag{5.54c}$$

---

[12] In particular, this shows that in the case $\mathrm{d}\tau = 0$, the Bargmann form obtained in part (iii) is the same as that constructed in Remark 5.12.

[13] Note that in particular we recover one direction of Theorem 3.50 (iv): if the Levi-Civita connection, with $\overset{\text{L}}{T} = 0$, has a regular limit, then $\mathrm{d}\tau = 0$.

Proof (i) From the duality conditions

$$1 = E^0(E_0) = \tau(v) + O(c^{-2}), \tag{5.55a}$$

$$0 = E^0(E_a) = c\tau(e_a) + O(c^{-1}), \tag{5.55b}$$

$$0 = E^a(E_0) = c^{-1}e^a(v) + O(c^{-2}), \tag{5.55c}$$

$$\delta^a_b = E^a(E_b) = e^a(e_b) + O(c^{-1}), \tag{5.55d}$$

we obtain that (v, e^a) and (τ, e_a) are a local frame of vector fields and the corresponding dual frame of one-forms, respectively. Inserting the frame expansions into the metric and inverse metric expressed in terms of the frames, we further obtain

$$\begin{aligned}
g &= \eta_{AB}E^A \otimes E^B = -E^0 \otimes E^0 + \delta_{ab}E^a \otimes E^b \\
&= -c^2\tau \otimes \tau - \tau \otimes a - a \otimes \tau + \delta_{ab}e^a \otimes e^b + O(c^{-1}), \tag{5.56a} \\
g^{-1} &= \eta^{AB}E_A \otimes E_B = -E_0 \otimes E_0 + \delta^{ab}E_a \otimes E_b \\
&= \delta^{ab}e_a \otimes e_b + O(c^{-2}). \tag{5.56b}
\end{aligned}$$

Thus, we see that the assumptions of Lemma 3.48 are satisfied with the given τ and $h = \delta^{ab}e_a \otimes e_b$.

(ii) The above equation for h in terms of the frame (v, e_a) and the fact that $\tau(v) = 1$ show that this frame is a Galilei frame.

(iii) Transforming the frame and dual frame by a local boost Λ parametrised as in (5.49), we obtain

$$\begin{aligned}
c\tilde{\tau} + c^{-1}\tilde{a} + O(c^{-2}) = \tilde{E}^0 &= \Lambda^0{}_A E^A \\
&= \Lambda^0{}_0 E^0 + \Lambda^0{}_a E^a \\
&= \left(1 + c^{-2}\frac{|k|^2}{2} + O(c^{-4})\right)(c\tau + c^{-1}a + O(c^{-2})) \\
&\quad + (c^{-1}k_a + O(c^{-3}))(e^a + O(c^{-1})) \\
&= c\tau + c^{-1}\left(a + \frac{|k|^2}{2}\tau + k_a e^a\right) + O(c^{-2}), \tag{5.57a} \\
\tilde{e}^a + O(c^{-1}) = \tilde{E}^a &= \Lambda^a{}_B E^B \\
&= \Lambda^a{}_0 E^0 + \Lambda^a{}_b E^b \\
&= (c^{-1}k^a + O(c^{-3}))(c\tau + c^{-1}a + O(c^{-2})) \\
&\quad + \left(\delta^a_b + c^{-2}\frac{k^a k_b}{2} + O(c^{-4})\right)(e^b + O(c^{-1})) \\
&= k^a\tau + e^a + O(c^{-1}), \tag{5.57b} \\
c^{-1}\tilde{v} + O(c^{-2}) = \tilde{E}_0 &= (\Lambda^{-1})^A{}_0 E_A \\
&= (\Lambda^{-1})^0{}_0 E_0 + (\Lambda^{-1})^a{}_0 E_a
\end{aligned}$$

$$= \left(1 + c^{-2}\frac{|k|^2}{2} + O(c^{-4})\right)(c^{-1}v + O(c^{-2}))$$
$$+ (-c^{-1}k^a + O(c^{-3}))(e_a + O(c^{-2}))$$
$$= c^{-1}(v - k^a e_a) + O(c^{-2}), \tag{5.57c}$$

$$\tilde{e}_a + O(c^{-2}) = \tilde{E}_a = (\Lambda^{-1})^B{}_a E_B$$
$$= (\Lambda^{-1})^0{}_a E_0 + (\Lambda^{-1})^b{}_a E_b$$
$$= (-c^{-1}k_a + O(c^{-3}))(c^{-1}v + O(c^{-2}))$$
$$+ \left(\delta_a^b + c^{-2}\frac{k^b k_a}{2} + O(c^{-4})\right)(e_b + O(c^{-2}))$$
$$= e_a + O(c^{-2}). \tag{5.57d}$$

From this, we may read off the transformation behaviour

$$(v, e_a) \mapsto (v - k^a e_a, e_a) = (v, e_a) \cdot (\mathbb{1}, k)^{-1}, \tag{5.58a}$$
$$(\tau, e^a) \mapsto (\tau, e^a + k^a \tau), \tag{5.58b}$$
$$a \mapsto a + k_a e^a + \frac{1}{2}|k|^2 \tau, \tag{5.58c}$$

which is the correct transformation behaviour for local Galilei frames, their dual frames, and the local representatives of Bargmann forms under local Galilei boosts, according to Proposition 4.16 and Construction 5.5.

(iv) In addition to the frame and dual frame expansions (5.48), we now denote the order-c^{-2} term of E_a by π_a, i.e. write

$$E_a = e_a + c^{-2}\pi_a + O(c^{-3}). \tag{5.59}$$

Duality then implies

$$0 = E^0(E_a) = c\tau(e_a) + c^{-1}(\tau(\pi_a) + a(e_a)) + O(c^{-2}). \tag{5.60}$$

In particular, this shows that

$$\tau(\pi_a) = -a(e_a). \tag{5.61}$$

Using (5.59), for the expansion of the inverse metric we obtain

$$g^{-1} = -E_0 \otimes E_0 + \delta^{ab} E_a \otimes E_b$$
$$= \delta^{ab} e_a \otimes e_b + c^{-2}(-v \otimes v + \delta^{ab}(e_a \otimes \pi_b + \pi_a \otimes e_b)) + O(c^{-3}). \tag{5.62}$$

Comparing this to $g^{-1} = h + c^{-2}m + O(c^{-3})$, we can read off $m = -v \otimes v + \delta^{ab}(e_a \otimes \pi_b + \pi_a \otimes e_b)$. Therefore, the natural unit timelike vector field from the metric expansion is given by

$$\hat{v} = -m(\tau, \cdot) = v - \delta^{ab}\tau(\pi_a) \otimes e_b , \tag{5.63}$$

where we used that $\tau(e_a) = 0$. Combined with (5.61), we obtain

$$\hat{v} = v + \delta^{ab}a(e_a) \otimes e_b = v + h(a, \cdot). \tag{5.64}$$

This is precisely the definition of the boost-invariant unit timelike vector field determined by the Bargmann form $\boldsymbol{a}$ according to Construction 5.15.

The function $\hat{\phi}$ determined by the metric expansion is given by $\hat{\phi} = -\frac{1}{2}g^{(0)}(\hat{v}, \hat{v})$, where $g = -c^2\tau \otimes \tau + g^{(0)} + O(c^{-1})$. By (5.56a), we have $g^{(0)} = -\tau \otimes a - a \otimes \tau + \delta_{ab}e^a \otimes e^b$. Hence, using (5.64) we obtain

$$\begin{aligned}
\hat{\phi} &= -\frac{1}{2}g^{(0)}(\hat{v}, \hat{v}) \\
&= a(\hat{v}) - \frac{1}{2}\delta_{ab}e^a(\hat{v})e^b(\hat{v}) \\
&= a(v) + h(a, a) - \frac{1}{2}\delta_{ab}h(a, e^a)h(a, e^b) \\
&= a(v) + \frac{1}{2}h(a, a). \tag{5.65}
\end{aligned}$$

By Construction 5.15, this is the function determined by $\hat{a} = \hat{\phi}\tau$, where $\hat{a}$ is the local representative of $\boldsymbol{a}$ with respect to $\hat{v}$.

(v) The local connection form $(\overset{L}{\hat{\omega}}{}^A{}_B)$ of $\overset{L}{\nabla}$ with respect to the local orthonormal frame (E_A) takes values in the Lorentz algebra, i.e. we have

$$\overset{L}{\hat{\omega}}{}^0{}_0 = 0, \quad \overset{L}{\hat{\omega}}{}^a{}_0 = \delta^{ab}\overset{L}{\hat{\omega}}{}^0{}_b , \quad (\overset{L}{\hat{\omega}}{}^a{}_b) \in \mathfrak{so}(n). \tag{5.66}$$

According to the general behaviour of local connection forms under frame changes, $(\overset{L}{\hat{\omega}}{}^A{}_B)$ can be expressed in terms of the local connection form $(\overset{L}{\omega}{}^A{}_B)$ of $\overset{L}{\nabla}$ with respect to the frame (v, e_a) as

$$\overset{L}{\hat{\omega}}{}^A{}_B = (X^{-1})^A{}_C \, \overset{L}{\omega}{}^C{}_D \, X^D{}_B + (X^{-1})^A{}_C \, dX^C{}_B \tag{5.67a}$$

with

$$E_A = X^B{}_A e_B . \tag{5.67b}$$

From the frame and dual frame expansions, we obtain the frame change matrix and its inverse as

$$X^t{}_0 = \tau(\mathrm{E}_0) = c^{-1} + \mathrm{O}(c^{-2}), \qquad X^a{}_0 = e^a(\mathrm{E}_0) = \mathrm{O}(c^{-2}), \quad (5.68a)$$

$$X^t{}_a = \tau(\mathrm{E}_a) = \mathrm{O}(c^{-2}), \qquad X^a{}_b = e^a(\mathrm{E}_b) = \delta^a_b + \mathrm{O}(c^{-2}), \quad (5.68b)$$

$$(X^{-1})^0{}_t = \mathrm{E}^0(v) = c + \mathrm{O}(c^{-1}), \qquad (X^{-1})^a{}_t = \mathrm{E}^a(v) = \mathrm{O}(c^{-1}), \quad (5.68c)$$

$$(X^{-1})^0{}_a = \mathrm{E}^0(e_a) = \mathrm{O}(c^{-1}), \qquad (X^{-1})^a{}_b = \mathrm{E}^a(e_b) = \delta^a_b + \mathrm{O}(c^{-1}). \quad (5.68d)$$

Together with the assumption that the $\widehat{\overset{\mathrm{L}}{\omega}}{}^A{}_B$ have regular formal $c \to \infty$ limits, which means that they are of the order c^0, we thus obtain

$$\overset{\mathrm{L}}{\omega}{}^a{}_0 = \underbrace{(X^{-1})^a{}_C \,\widehat{\overset{\mathrm{L}}{\omega}}{}^C{}_D\, X^D{}_0}_{=c^{-1}\widehat{\overset{\mathrm{L}}{\omega}}{}^a{}_t + \mathrm{O}(c^{-2})} + (X^{-1})^a{}_C \underbrace{\mathrm{d}X^C{}_0}_{=\mathrm{O}(c^{-2})} = c^{-1}\widehat{\overset{\mathrm{L}}{\omega}}{}^a{}_t + \mathrm{O}(c^{-2}), \quad (5.69a)$$

$$\overset{\mathrm{L}}{\omega}{}^a{}_b = \underbrace{(X^{-1})^a{}_C \,\widehat{\overset{\mathrm{L}}{\omega}}{}^C{}_D\, X^D{}_b}_{=\widehat{\overset{\mathrm{L}}{\omega}}{}^a{}_b + \mathrm{O}(c^{-1})} + (X^{-1})^a{}_C \underbrace{\mathrm{d}X^C{}_b}_{=\mathrm{O}(c^{-2})} = \widehat{\overset{\mathrm{L}}{\omega}}{}^a{}_b + \mathrm{O}(c^{-1}), \quad (5.69b)$$

which proves an expansion as in (5.52).

A direct calculation as in the proof of (iii) shows that the local forms $(\omega^a{}_b, \varpi^a)$ transform under local Galilei boosts of the frame as the local connection form of a Galilei connection would. One easily sees that the same is true for spatial rotations since those are c-independent and thus rotations of (E_a) directly give rotations of (e_a). Therefore, we obtain a Galilei connection ∇.

From the expansions of the dual frame (5.48) and the connection form (5.52) we obtain the torsion of $\overset{\mathrm{L}}{\nabla}$ using Cartan's first structure equation as

$$\begin{aligned}
\overset{\mathrm{L}}{T}{}^0 &= \mathrm{dE}^0 + \overset{\mathrm{L}}{\omega}{}^0{}_A \wedge \mathrm{E}^A \\
&= c\,\mathrm{d}\tau + c^{-1}\mathrm{d}a + c^{-1}\varpi_a \wedge e^a + \mathrm{O}(c^{-2}) \\
&= c\,\mathrm{d}\tau + c^{-1}f + \mathrm{O}(c^{-2}), \quad (5.70a) \\
\overset{\mathrm{L}}{T}{}^a &= \mathrm{dE}^a + \overset{\mathrm{L}}{\omega}{}^a{}_B \wedge \mathrm{E}^B \\
&= \mathrm{d}e^a + \varpi^a\tau + \omega^a{}_b \wedge e^b + \mathrm{O}(c^{-1}) \\
&= T^a + \mathrm{O}(c^{-1}). \quad (5.70b)
\end{aligned}$$

Finally, inserting the expansion of the connection form (5.52) into Cartan's second structure equation for the curvature form,

$$\overset{(\mathrm{L})}{R}{}^A{}_B = \mathrm{d}\overset{(\mathrm{L})}{\omega}{}^A{}_B + \overset{(\mathrm{L})}{\omega}{}^A{}_C \wedge \overset{(\mathrm{L})}{\omega}{}^C{}_B , \quad (5.71)$$

a short calculation yields (5.54). $\qquad\qquad\qquad\qquad\qquad\qquad\qquad\square$

Exercise 5.26 (*The Newtonian limit of the* NUT *region via Bargmann forms*) In this exercise, we are going to use Theorem 5.25 to compute the Newton–Cartan limit of the NUT region of Taub–NUT spacetime, and show that it doesn't have absolute rotation—as we did in Exercise 3.55—in a very efficient way.

The Taub–NUT metric is

$$g = U(c\,\mathrm{d}t - 4l\sin^2(\tfrac{\theta}{2})\,\mathrm{d}\varphi)^2 - \frac{1}{U}\mathrm{d}r^2 + (r^2 + l^2)(\mathrm{d}\theta^2 + \sin^2\theta\,\mathrm{d}\varphi^2), \quad (5.72\mathrm{a})$$

where $U = -1 + 2\big(\frac{GM}{c^2}r + l^2\big)/(r^2 + l^2)$. The inverse metric is given by

$$g^{-1} = \left(\frac{1}{U} + \frac{4l^2\tan^2(\tfrac{\theta}{2})}{r^2 + l^2}\right)c^{-2}\partial_t \otimes \partial_t + \frac{l/\cos^2(\tfrac{\theta}{2})}{r^2 + l^2}c^{-1}(\partial_t \otimes \partial_\varphi + \partial_\varphi \otimes \partial_t)$$

$$- U\,\partial_r \otimes \partial_r + \frac{1}{r^2 + l^2}(\partial_\theta \otimes \partial_\theta + \sin^{-2}\theta\,\partial_\varphi \otimes \partial_\varphi). \quad (5.72\mathrm{b})$$

We parametrise $l = \frac{J}{Mc}$, and consider the positive r NUT region $U < 0$.

(a) From the Taub–NUT metric, read off an orthonormal dual frame $(\mathrm{E}^0, \mathrm{E}^a)$ of one-forms. Compute the frame $(\mathrm{E}_0, \mathrm{E}_a) = (-(\mathrm{E}^0)^\sharp, (\mathrm{E}^a)^\sharp)$, where a 'sharp' sign denotes the metric-induced isomorphism sending dual vectors to vectors, i.e. $\alpha^\sharp = g^{-1}(\alpha, \cdot)$.
Hint: Since the metric is written as a sum of squares, you can directly read off the dual frame.

(b) By expanding the frame and dual frame as formal power series in c^{-1} and comparing to (5.48), determine the induced Galilei frame and dual frame for the limiting Galilei manifold, as well as the local representative of the induced Bargmann form.
Hint: Don't forget to insert $l = \frac{J}{Mc}$. For inverting formal power series, use the geometric series. The resulting spacelike frame and dual frame should simply be the standard orthonormal frame and dual frame for Euclidean $\mathbb{R}^3$ induced by spherical coordinates, $\tau = \mathrm{d}t$, and $v = \partial_t$.

(c) Argue that the resulting v is rigid, i.e. that $\mathcal{L}_v h = 0$.
Hint: Using $h = \delta^{ab}\mathrm{e}_a \otimes \mathrm{e}_b$, this is almost obvious, given the resulting v and e_a.

(d) According to (5.53), the Galilei connection that arises as the $c \to \infty$ limit of the Lorentzian Levi-Civita connection has vanishing extended torsion. Using this as well as the relationship between the Newton–Coriolis form and the local representatives of the Bargmann form and the mass torsion (Corollary 5.9), compute the Newton–Coriolis form of the limiting Galilei connection with respect to v. Comparing to the decomposition $\Omega = \tau \wedge \alpha + 2\omega$, read off the twist ω. Expressing the twist in terms of the spatial orthonormal coframe (e^a), read off the metric length of ω and observe that it is spatially non-constant, i.e. that we don't have absolute rotation.

Using the Bargmann form framework, we can thus compute the Newton–Coriolis form of a limiting Galilei manifold with respect to a *chosen* unit timelike vector field v far more easily than just using expansions of the metric (compare the amount of calculation required in this exercise to that from Exercise 3.55): we just have to find a Lorentzian orthonormal frame for which the expansion of the timelike vector field E_0 starts with the chosen v, and then expand. (For expansions of the metric, one directly obtains the Newton–Coriolis form with respect to the natural unit timelike vector field $\hat{v}$ that is determined by the expansion, such that one cannot specify this beforehand.)

$\triangle$

Exercise 5.27 (*Newtonian limits of matter actions*) In this exercise, we are going to show how to obtain the point particle and Schrödinger field actions from Sect. 5.4 as formal Newtonian limits of Lorentzian counterparts, using Theorem 5.25.

(a) We consider the action for the timelike future-directed worldline γ of a massive test particle in Lorentzian spacetime,

$$S_{\text{Lor.}}[\gamma] = -mc \int_{\lambda_i}^{\lambda_f} \mathrm{d}\lambda \sqrt{-g(\gamma', \gamma')}\,. \tag{5.73}$$

Show that subtracting the 'rest energy' term $S_{\text{rest}}[\gamma] = -mc^2 \int_\gamma \tau$, in the limit $c \to \infty$ this action goes over to the action for a massive point particle in a Galilei manifold that was introduced in Construction 5.21.
Hint: Insert $g = \eta_{AB} E^A \otimes E^B = -E^0 \otimes E^0 + \delta_{ab} E^a \otimes E^b$, use (5.48), and expand the action using the power series expansion of $\sqrt{1+x}$.

(b) Consider the action for a Klein–Gordon field Φ on Lorentzian spacetime,

$$S_{\text{KG}}[\Phi] = -c^{-1} \int \text{vol}_g \left(\hbar^2 g^{\mu\nu} \overline{\partial_\mu \Phi}\, \partial_\nu \Psi + m^2 c^2 |\Phi|^2 \right)\,. \tag{5.74}$$

Assuming that $\tau = \mathrm{d}t$ is exact and separating off an oscillating 'rest energy' phase factor with positive frequency from the Klein–Gordon field according to

$$\Phi = \mathrm{e}^{-\mathrm{i}\frac{mc^2}{\hbar^2}t}\Psi, \tag{5.75}$$

show that in the limit $c \to \infty$ the Klein–Gordon action goes over to the Schrödinger field action in a Galilei spacetime for Ψ, as introduced in Construction 5.22.
Hint: First show that $\partial_\mu \Phi = \mathrm{e}^{-\mathrm{i}\frac{mc^2}{\hbar^2}t}(-\mathrm{i}\frac{mc^2}{\hbar}\tau_\mu + \partial_\mu)\Psi$. Then write the metric in terms of an orthonormal frame, which we expand as in (5.48). Note that you can compute the expression $cE_0^\mu \tau_\mu$ to order c^{-2} by using the dual basis condition $1 = E_0^\mu E_\mu^0$, and similarly $E_a^\mu \tau_\mu$ to c^{-2} using $0 = E_a^\mu E_\mu^0$.

$\triangle$

Construction 5.28 We now want to analyse how the formal $c \to \infty$ limit from Lorentzian to Galilei manifolds interacts with (active) transformations of the geometric objects by diffeomorphisms. With notations as in Theorem 5.25, we let $\varphi \colon M \to M$ be a diffeomorphism, and consider its action on all the Lorentzian objects by pushforward, i.e.

$$g \mapsto \varphi_* g, \quad \mathrm{E}_A \mapsto \varphi_* \mathrm{E}_A, \quad \overset{\mathrm{L}}{\nabla} \mapsto \varphi_* \overset{\mathrm{L}}{\nabla}. \tag{5.76a}$$

Note that the action of φ on the connection $\overset{\mathrm{L}}{\nabla}$ may also be described either by pushforward of local connection forms on $U \subset M$, or by pushforward of the global connection form on the linear frame bundle $F(M)$ by the natural lift of φ to $F(M)$. By duality, it is clear that the dual frame transforms as

$$\mathrm{E}^A \mapsto \varphi_* \mathrm{E}^A. \tag{5.76b}$$

Now directly inserting the formal expansions in powers of c^{-1}, we see that the 'limiting' objects for our Galilei manifold also transform by pushforward, i.e. as

$$\tau \mapsto \varphi_* \tau, \qquad h \mapsto \varphi_* h, \qquad e^a \mapsto \varphi_* e^a, \tag{5.77a}$$

$$v \mapsto \varphi_* v, \qquad e_a \mapsto \varphi_* e_a, \tag{5.77b}$$

$$a \mapsto \varphi_* a, \qquad \nabla \mapsto \varphi_* \nabla. \tag{5.77c}$$

Note that according to Construction 5.14, this means that the Bargmann form transforms as

$$\boldsymbol{a} \mapsto \boldsymbol{\varphi_* a}, \tag{5.77d}$$

where $\boldsymbol{\varphi} \colon G(M, \tau, h) \overset{\cong}{\to} G(M, \varphi_* \tau, \varphi_* h)$ is the natural lift of φ to an isomorphism of Galilei frame bundles.

Now, instead of considering transformations by proper diffeomorphisms of M, we want to transform the fields by 'c-dependent diffeomorphisms'. Of course, strictly speaking this does not make mathematical sense, since coordinate representations of diffeomorphisms are $\mathbb{R}^{n+1}$-valued, and thus cannot be formal power series in c^{-1}. However, when considering a family of diffeomorphisms φ_ε smoothly depending on a proper small parameter ε with $\varphi_0 = \mathrm{id}_M$, its pushforward action on any natural geometric object A may be expanded using the Lie derivative as

$$(\varphi_\varepsilon)_* A = A - \varepsilon \mathcal{L}_X A + \mathrm{O}(\varepsilon^2), \tag{5.78}$$

where X is the vector field generating of the family of diffeomorphisms. Therefore, the way to properly implement 'pushforward by c-dependent diffeomorphisms close to the identity' is to consider the action

$$A \mapsto A - c^{-2} \mathcal{L}_X A + \mathrm{O}(c^{-4}) \tag{5.79a}$$

for a vector field X, which is thought of as being the leading order expansion of action with the pushforward of the diffeomorphism $\exp(c^{-2}X)$.[14] Now applying this action to some object A expanded as

$$A = c^{-k}(A_{(0)} + c^{-1}A_{(1)} + c^{-2}A_{(2)} + \mathrm{O}(c^{-3})) , \qquad (5.79b)$$

we obtain

$$A \mapsto c^{-k}\left(A_{(0)} + c^{-1}A_{(1)} + c^{-2}(A_{(2)} - \mathcal{L}_X A_{(0)}) + \mathrm{O}(c^{-4})\right): \qquad (5.79c)$$

the leading coefficient stays invariant, and only starting at the next order coefficient we obtain a non-trivial transformation. Applying this to all our Lorentzian objects, for the 'limiting' Galilei-manifold objects the only non-trivial transformations we obtain are

$$a \mapsto a - \mathcal{L}_X \tau = a - \mathrm{d}\tau(X, \cdot) - \mathrm{d}(\tau(X)), \qquad (5.80a)$$

$$f \mapsto f - \mathcal{L}_X \mathrm{d}\tau = f - \mathrm{d}(\mathrm{d}\tau(X, \cdot)), \qquad (5.80b)$$

where we used Cartan's magic formula. (Of course, the transformation (5.80b) of the local representative of the mass torsion also follows from that of the Bargmann form (5.80a) by using that, according to Corollary 5.9, $f = \mathrm{d}a + \Omega$.)

In Lorentzian geometry, acting on all geometric objects involved in some discussion with pushforward by a diffeomorphism leads to a different mathematical description of what is to be considered 'the same' geometric/physical situation.[15] Therefore, when viewing Galilei geometry with a Bargmann form as arising as the Newtonian limit of Lorentzian geometry, we need to consider the transformation (5.80a), for any vector field X, as a transformation relating two equivalent descriptions of the same geometric situation. These transformations (5.80a), arising from subleading c^{-2}-dependent diffeomorphisms in Lorentzian geometry, are called *type* II *gauge transformations* of Bargmann forms. $\triangle$

Remark 5.29 In the case of absolute time, i.e. $\mathrm{d}\tau = 0$, the type II gauge transformation (5.80a) amounts to

$$a \mapsto a - \mathrm{d}\tau(X, \cdot) - \mathrm{d}(\tau(X)) = a - \mathrm{d}(\tau(X)), \qquad (5.81)$$

i.e. to a type I gauge transformation (5.25) with transformation parameter $\chi = -\tau(X)$. So in the case $\mathrm{d}\tau = 0$, both types of gauge transformations agree.

---

[14] Considering $c^{-1}X$ instead would spoil the expansions assumed in Theorem 5.25.

[15] Of course, this idea is true more generally: in any differential-geometric discussion involving only *natural* geometric objects on some manifold, i.e. objects on which pushforward by any diffeomorphism may be defined in a natural way, acting on *all* involved objects with a diffeomorphism leads to an equivalent situation.

In the case $d\tau \neq 0$, however, they are different. Hence, when viewing Galilei geometry with a Bargmann form as arising as the Newtonian limit of Lorentzian geometry, the type I transformations have no natural interpretation as encoding 'redundancies' in the description of the geometric situation. This shows that the description of Bargmann forms in terms of connections on a **Barg** extension of the Galilei frame bundle, as discussed in Section 5.3, is not the natural setting for the description of Newtonian limits, at least when allowing for non-absolute time. This was first discussed in the articles [8,9].   △

5.6   Null Reduction of Lorentzian Geometry

In this section, we will discuss how the quotient of a Lorentzian manifold by a null (i.e. lightlike) symmetry naturally carries the structure of a Galilei manifold, and how a minimal additional choice of structure endows it with a Bargmann form. This procedure is called *null reduction*, and is well-known in the literature [3,10]. However, differently from most literature, our discussion will be in fully coordinate-free terms.

Definition 5.30 A *null-symmetric Lorentzian manifold* is a Lorentzian manifold $(\tilde{M}, \tilde{g})$ with a nowhere-vanishing null Killing field ξ, i.e. a nowhere-vanishing vector field $\xi \in \Gamma(T\tilde{M})$ satisfying $\mathcal{L}_\xi \tilde{g} = 0$ and $\tilde{g}(\xi, \xi) = 0$. We write $\dim \tilde{M} = n + 2$, assuming that $n \geq 1$.

We will assume that ξ is complete, i.e. that its flow lines can be extended to the whole of $\mathbb{R}$. We will further assume that the flow lines of ξ are not closed (which follows, e.g., if we assume that $(\tilde{M}, \tilde{g})$ have no closed causal curves), such that the action of $\mathbb{R}$ on $\tilde{M}$ by the flow of ξ is free. Finally, we assume that the flow action of $\mathbb{R}$ on $\tilde{M}$ is proper[16], such that the quotient $M := \tilde{M} \Big/ \text{(flow of } \xi)$ is a (Hausdorff) manifold. Then $\tilde{M} \xrightarrow{\pi} M$ is a principal $\mathbb{R}$-bundle (with action given by the flow of ξ).   ✿

In the following, let $(\tilde{M}, \tilde{g}, \xi)$ be a null-symmetric Lorentzian manifold and $\tilde{M} \xrightarrow{\pi} M$ its quotient by the flow of ξ. We will several times need the following result:

Lemma 5.31 *Let $\tilde{\alpha} \in \Omega^1(\tilde{M})$ be a one-form that is invariant under the flow of ξ and satisfies $\tilde{\alpha}(\xi) = 0$. Then there is a unique one-form $\alpha \in \Omega^1(M)$ satisfying*

$$\pi^*\alpha = \tilde{\alpha}. \tag{5.82}$$

Proof An α as desired would need to satisfy $\tilde{\alpha}(\tilde{w}) = (\pi^*\alpha)(\tilde{w}) = \alpha|_{\pi(q)}(\mathrm{D}\pi(\tilde{w}))$ for any $\tilde{w} \in T_q\tilde{M}$. Hence, given $w \in T_pM$, we need to define $\alpha(w)$ as $\alpha(w) := \tilde{\alpha}(\tilde{w})$,

---

[16] The action being proper means that $\mathbb{R} \times \tilde{M} \ni (t, q) \mapsto (q, \Phi_t(q)) \in \tilde{M} \times \tilde{M}$ is a proper map, where Φ denotes the action / flow map.

for some $q \in \tilde{M}$ with $\pi(q) = p$ and some $\tilde{w} \in T_q M$ with $D\pi(\tilde{w}) = w$. We need to prove that this is well-defined, i.e. independent of the choices of $\tilde{w}$ and q.

Independence of the choice of $\tilde{w}$ (fixing q) follows easily: any two choices of $\tilde{w} \in T_q \tilde{M}$ with $D\pi(\tilde{w}) = w$ differ by an element of $\ker D\pi|_q = \mathrm{span}\{\xi|_q\}$, and by assumption $\tilde{\alpha}$ vanishes on $\mathrm{span}\{\xi\}$.

Further, using invariance of $\tilde{\alpha}$ under the flow of ξ, we may show that $\tilde{\alpha}(\tilde{w})$ is independent of the choice of q as well: any two choices q and q_2 are related by the flow, i.e. there is a $t \in \mathbb{R}$ such that $q_2 = \Phi_t(q)$, where Φ denotes the flow. Choosing corresponding vectors $\tilde{w} \in T_q \tilde{M}$, $\tilde{w}_2 \in T_{q_2} \tilde{M}$ with $D\pi(\tilde{w}) = w = D\pi(\tilde{w}_2)$, the invariance of $\tilde{\alpha}$ implies

$$\tilde{\alpha}|_q(\tilde{w}) = (\Phi_t^* \tilde{\alpha})|_q(\tilde{w}) = \tilde{\alpha}|_{\Phi_t(q)}(D\Phi_t(\tilde{w})) = \tilde{\alpha}|_{q_2}(D\Phi_t(\tilde{w})) = \tilde{\alpha}|_{q_2}(\tilde{w}_2),$$
$$(5.83)$$

where in the last step we used that both $D\Phi_t(\tilde{w})$ and $\tilde{w}_2$ are vectors in $T_{q_2} \tilde{M}$ projecting to w under $D\pi$, and we already showed independence of the choice of such a vector. $\square$

Proposition 5.32 *(i) There is a unique one-form $\tau \in \Omega^1(M)$ satisfying $\pi^* \tau = \tilde{g}(\xi, \cdot)$.*

(ii) There is a unique symmetric contravariant 2-tensor field $h \in \Gamma(\bigvee^2 TM)$ satisfying $\tilde{g}^{-1}(\pi^* \alpha, \pi^* \beta) = h(\alpha, \beta) \circ \pi$ for all $\alpha, \beta \in \Omega^1(M)$.

(iii) (M, τ, h) is a Galilei manifold.

Proof (i) Since ξ is a Killing field, we have $\mathcal{L}_\xi(\tilde{g}(\xi, \cdot)) = (\mathcal{L}_\xi \tilde{g})(\xi, \cdot) + \tilde{g}(\mathcal{L}_\xi \xi, \cdot) = 0$. This means that the one-form $\tilde{g}(\xi, \cdot) \in \Omega^1(\tilde{M})$ is invariant under the flow of ξ. Further, since ξ is null, $\tilde{g}(\xi, \cdot)$ vanishes on ξ. Therefore, existence of a unique τ as desired follows by Lemma 5.31.

(ii) Denoting the flow of ξ by Φ, by definition we have $\pi \circ \Phi_t = \pi$ (the fibres of π are the flow lines, which the flow leaves invariant). Using this, for any one-form $\alpha \in \Omega^1(M)$ we have $\Phi_t^*(\pi^* \alpha) = (\pi \circ \Phi_t)^* \alpha = \pi^* \alpha$, i.e. the pullback $\pi^* \alpha \in \Omega^1(\tilde{M})$ is invariant under the flow of ξ.

Hence, given two one-forms $\alpha, \beta \in \Omega^1(M)$, the function $\tilde{g}^{-1}(\pi^* \alpha, \pi^* \beta) \in C^\infty(\tilde{M})$ is invariant under the flow of ξ (using that ξ is a Killing field). Therefore, we can define the function $h(\alpha, \beta) \in C^\infty(M)$ by $\tilde{g}^{-1}(\pi^* \alpha, \pi^* \beta) = h(\alpha, \beta) \circ \pi$.

This defines a map $\Omega^1(M) \times \Omega^1(M) \ni (\alpha, \beta) \mapsto h(\alpha, \beta) \in C^\infty(M)$. Obviously, this is $\mathbb{R}$-linear in both its arguments and symmetric. By direct calculation, we also see that it is $C^\infty(M)$-linear in both arguments: for $\alpha, \beta \in \Omega^1(M)$ and

$f \in C^\infty(M)$ we have

$$
\begin{aligned}
h(f\alpha, \beta) \circ \pi &= \tilde{g}^{-1}(\pi^*(f\alpha), \pi^*\beta) \\
&= \tilde{g}^{-1}((f \circ \pi) \cdot \pi^*\alpha, \pi^*\beta) \\
&= (f \circ \pi) \cdot \tilde{g}^{-1}(\pi^*\alpha, \pi^*\beta) \\
&= (f \circ \pi) \cdot (h(\alpha, \beta) \circ \pi) \\
&= (f h(\alpha, \beta)) \circ \pi,
\end{aligned}
\tag{5.84}
$$

implying $h(f\alpha, \beta) = f h(\alpha, \beta)$; linearity in its second argument follows analogously or by symmetry.

Thus we obtain a tensor field h as stated.

(iii) Since ξ is nowhere-vanishing, $\tilde{g}(\xi, \cdot)$ is nowhere-vanishing. Hence τ, defined by $\pi^*\tau = \tilde{g}(\xi, \cdot)$, is nowhere-vanishing as well.

We now have to show that h is positive semidefinite of rank n, with degenerate direction spanned by τ. In order to show this, consider a point $p \in M$, and choose a point $q \in \tilde{M}$ with $\pi(q) = p$. The differential $\mathrm{D}\pi|_q : T_q\tilde{M} \twoheadrightarrow T_pM$ is surjective, hence its dual map $(\mathrm{D}\pi|_q)^* : T_p^*M \hookrightarrow T_q^*\tilde{M}, \alpha \mapsto \alpha \circ \mathrm{D}\pi|_q$ is injective. Evaluating the defining equation for h at q, we see that $h|_p$ can be written as

$$
h|_p(\alpha, \beta) = \tilde{g}^{-1}|_q((\mathrm{D}\pi|_q)^*(\alpha), (\mathrm{D}\pi|_q)^*(\beta)) \text{ for } \alpha, \beta \in T_p^*M.
\tag{5.85}
$$

Put differently, we obtain $h|_p$ by restricting $\tilde{g}^{-1}|_q$ to $\mathrm{im}\,(\mathrm{D}\pi|_q)^* \subset T_q^*\tilde{M}$ and identifying $\mathrm{im}\,(\mathrm{D}\pi|_q)^*$ with T_p^*M via $(\mathrm{D}\pi|_q)^*$. Note also that by definition of τ, we have $(\mathrm{D}\pi|_q)^*(\tau|_p) = \tilde{g}|_q(\xi|_q, \cdot)$.

So we need to show that the restriction of $\tilde{g}^{-1}|_q$ to $\mathrm{im}\,(\mathrm{D}\pi|_q)^*$ is positive semidefinite of rank n, with degenerate direction spanned by $\tilde{g}|_q(\xi|_q, \cdot) =: \eta$. Now for any $\alpha \in T_p^*M$, we have

$$
\tilde{g}^{-1}|_q(\eta, (\mathrm{D}\pi|_q)^*(\alpha)) = ((\mathrm{D}\pi|_q)^*(\alpha))(\xi|_q) = \alpha(\mathrm{D}\pi|_q(\xi|_q)) = 0.
\tag{5.86}
$$

This shows that $\mathrm{im}\,(\mathrm{D}\pi|_q)^*$ lies in the $\tilde{g}^{-1}|_q$-orthogonal complement of η; for dimensional reasons, it therefore *is* the orthogonal complement of η. But for any vector space with a Lorentzian-signature non-degenerate symmetric bilinear form, the restriction of the form to the orthogonal complement of a null vector is positive semidefinite with precisely one degenerate direction spanned by the null vector. Hence we are finished. $\qquad\square$

Now we are going to show that any vector field on M can be lifted to a vector field on $\tilde{M}$ that is invariant under the flow of ξ and projects down to the original one:

Proposition 5.33 *For any vector field $X \in \Gamma(TU)$ on an open set $U \subset M$, there is an invariant lift $\tilde{X} \in \Gamma(T(\pi^{-1}(U)))$ on $\pi^{-1}(U) \subset \tilde{M}$ satisfying $\mathcal{L}_\xi \tilde{X} = 0$ and $\mathrm{D}\pi \circ \tilde{X} = X \circ \pi$. It is unique up to addition of a term of the form $(\lambda \circ \pi) \cdot \xi$, where $\lambda \in C^\infty(U)$.*

Proof To show uniqueness, consider the difference A of two such lifts. This satisfies $\mathcal{L}_\xi A = 0$ and $\mathrm{D}\pi \circ A = 0$. Since π is the quotient map by the flow of ξ, we have $\ker \mathrm{D}\pi = \mathrm{span}\{\xi\}$. Therefore $\mathrm{D}\pi \circ A = 0$ implies $A = f\xi$ for some function f on $\tilde{M}$. From this, we obtain $0 = \mathcal{L}_\xi A = \mathcal{L}_\xi(f\xi) = \mathcal{L}_\xi f \cdot \xi + f \cdot \mathcal{L}_\xi \xi = \mathcal{L}_\xi f \cdot \xi$. Since ξ is nowhere-vanishing, we have $\mathcal{L}_\xi f = 0$. This means that f is invariant under the flow of ξ, i.e. constant along the flow lines; hence it is of the form $f = \lambda \circ \pi$ for a function λ on U.

For existence, note that it is enough to prove it locally: if we have an open cover $\{V_i\}_{i \in \mathcal{I}}$ of U such that for each V_i we have an invariant lift $\tilde{X}_i$ of X on $\pi^{-1}(V_i)$, using a partition of unity $\{\psi_i\}_{i \in \mathcal{I}}$ subordinate to $\{V_i\}_{i \in \mathcal{I}}$ we obtain an invariant lift on all of U as

$$\tilde{X} = \sum_{i \in \mathcal{I}} (\psi_i \circ \pi) \cdot \tilde{X}_i \, . \tag{5.87}$$

In particular, since $\tilde{M} \xrightarrow{\pi} M$ is a principal bundle, we can cover U by a family of open sets on which local sections of π exist. Thus we need only prove existence of an invariant lift for a vector field defined on an open set on which we have a local section of π.[17]

So without loss of generality let $\sigma : U \to \tilde{M}$ be a local section of π. On the image of U under σ, we define $\tilde{X}$ by $\tilde{X}|_{\sigma(p)} = \mathrm{D}\sigma(X|_p)$, such that it satisfies

$$\mathrm{D}\pi(\tilde{X}|_{\sigma(p)}) = \mathrm{D}\pi(\mathrm{D}\sigma(X|_p)) = (\mathrm{D}(\pi \circ \sigma))(X|_p)$$
$$= \mathrm{D}(\mathrm{id}_M)(X|_p) = X|_p = X|_{\pi(\sigma(p))} \tag{5.88}$$

as desired. The condition $\mathcal{L}_\xi \tilde{X} = 0$ then determines a unique extension of this field from $\sigma(U)$ to all of $\pi^{-1}(U)$. Explicitly, denoting the flow of ξ by Φ, it is given by

$$\tilde{X}|_{\Phi_t(\sigma(p))} = \mathrm{D}\Phi_t(\tilde{X}|_{\sigma(p)}). \tag{5.89}$$

Since by definition we have $\pi \circ \Phi_t = \pi$, this satisfies

$$\mathrm{D}\pi(\tilde{X}|_{\Phi_t(\sigma(p))}) = \mathrm{D}\pi(\mathrm{D}\Phi_t(\tilde{X}|_{\sigma(p)}))$$
$$= \mathrm{D}(\pi \circ \Phi_t)(\tilde{X}|_{\sigma(p)})$$
$$= \mathrm{D}\pi(\tilde{X}|_{\sigma(p)})$$
$$= X|_p$$
$$= X|_{\pi(\Phi_t(\sigma(p)))} \, , \tag{5.90}$$

showing that the extension along the flow projects to X as desired. $\square$

---

[17] In fact, any principal $\mathbb{R}$-bundle is *globally* trivial (see Remark 5.38). However, we want to present the argument for the existence of invariant lifts without using this fact.

Remark 5.34 For two vector fields X, Y on M and invariant lifts $\tilde{X}, \tilde{Y}$ thereof, by the Jacobi identity $[\tilde{X}, \tilde{Y}]$ is invariant under the flow of ξ as well.

Furthermore, it is a well-known fact that for any smooth map $\varphi \colon L \to N$ and vector fields $A_1, A_2 \in \Gamma(TL)$, $B_1, B_2 \in \Gamma(TN)$ the assumption that the A_i be φ-*related* to the B_i, i.e. $\mathrm{D}\varphi \circ A_i = B_i \circ \varphi$, implies that the commutators are φ-related as well, i.e. $\mathrm{D}\varphi \circ [A_1, A_2] = [B_1, B_2] \circ \varphi$. In the current situation, this implies that $\mathrm{D}\pi \circ [\tilde{X}, \tilde{Y}] = [X, Y] \circ \pi$.

Therefore, the vector field commutator of invariant lifts is an invariant lift of the commutator, or symbolically

$$[\tilde{X}, \tilde{Y}] = \widetilde{[X, Y]}. \tag{5.91}$$

$\triangle$

Proposition 5.35 *For X, Y spacelike vector fields on (M, τ, h), choosing any invariant lifts $\tilde{X}, \tilde{Y}$, we have $\tilde{g}(\tilde{X}, \tilde{Y}) = {}^{(n)}h(X, Y) \circ \pi$.*

Proof Consider the one-form $\tilde{g}(\tilde{X}, \cdot) \in \Omega^1(\tilde{M})$. Since both $\tilde{g}$ and $\tilde{X}$ are invariant under the flow of ξ, this form is invariant as well. Furthermore, by definition of τ and by X being spacelike we have

$$\tilde{g}(\tilde{X}, \xi) = (\pi^*\tau)(\tilde{X}) = \tau(X) \circ \pi = 0, \tag{5.92}$$

i.e. the form vanishes on ξ. Therefore, Lemma 5.31 implies that there is a unique one-form $\alpha \in \Omega^1(M)$ defined by

$$\pi^*\alpha = \tilde{g}(\tilde{X}, \cdot). \tag{5.93}$$

By definition of h, for any $\beta \in \Omega^1(M)$ this α satisfies

$$h(\alpha, \beta) \circ \pi = \tilde{g}^{-1}(\pi^*\alpha, \pi^*\beta) = (\pi^*\beta)(\tilde{X}) = \beta(X) \circ \pi. \tag{5.94}$$

This shows that $h(\alpha, \cdot) = X$. Therefore we have ${}^{(n)}h(X, Y) = \alpha(Y)$. Finally, this implies that ${}^{(n)}h(X, Y) \circ \pi = \alpha(Y) \circ \pi = (\pi^*\alpha)(\tilde{Y}) = \tilde{g}(\tilde{X}, \tilde{Y})$. $\qquad\square$

Proposition 5.36 *Let v be a unit timelike vector field on (M, τ, h).*

  (i) For any invariant lift $\tilde{v}$ of v, the one-form $\tilde{g}(\tilde{v}, \cdot) \in \Omega^1(\tilde{M})$ is a principal connection on $\tilde{M} \to M$.

  (ii) There is a unique invariant lift $\tilde{v}$ that is null, i.e. such that $\tilde{g}(\tilde{v}, \tilde{v}) = 0$.

Proof (i) In general, for a $\mathfrak{g}$-valued one-form $\omega \in \Omega^1(P, \mathfrak{g})$ on the total space P of a principal G-bundle to be a connection, we need it to be Ad-equivariant, i.e. satisfy

$$R_g^*\omega = \mathrm{Ad}_{g^{-1}} \circ \omega \text{ for all } g \in G, \tag{5.95a}$$

and to satisfy

$$\omega(\hat{X}) = X \text{ for all } X \in \mathfrak{g}, \tag{5.95b}$$

where $\hat{X} \in \Gamma(TP)$ denotes the fundamental vector field corresponding to the Lie algebra element X.

In the case at hand, since $\mathbb{R}$ is an abelian Lie group, Ad-equivariance simply amounts to invariance under the action. Further, ξ is the fundamental vector field of the flow action corresponding to the Lie algebra basis element $1 \in \mathbb{R}$. Therefore, for a one-form $\omega \in \Omega^1(\tilde{M})$ on $\tilde{M}$ to be a principal connection, we need it to be invariant under the flow of ξ, and to satisfy $\omega(\xi) = 1$.

For $\tilde{g}(\tilde{v}, \cdot)$, both of these hold immediately: on the one hand, $\tilde{g}$ and $\tilde{v}$ are invariant under the flow of ξ; on the other hand, by definition of τ and by v being unit timelike we have

$$\tilde{g}(\tilde{v}, \xi) = (\pi^*\tau)(\tilde{v}) = \tau(v) \circ \pi = 1. \tag{5.96}$$

(ii) Let $\hat{v}$ be an invariant lift of v that is not necessarily null. We know that any other invariant lift can be written as $\tilde{v} = \hat{v} + (\lambda \circ \pi) \cdot \xi$ with $\lambda \in C^\infty(M)$. For this to be null, we need

$$0 = \tilde{g}(\tilde{v}, \tilde{v}) = \tilde{g}(\hat{v}, \hat{v}) + 2(\lambda \circ \pi) \cdot \tilde{g}(\hat{v}, \xi) = \tilde{g}(\hat{v}, \hat{v}) + 2\lambda \circ \pi, \tag{5.97}$$

which uniquely fixes λ. $\qquad\square$

Remark 5.37 The proof of Proposition 5.36 (i) shows that conversely for any principal connection ω on $\tilde{M}$, the invariant field $\tilde{g}^{-1}(\omega, \cdot)$ projects to a unit timelike vector field on (M, τ, h). $\qquad\triangle$

Remark 5.38 Using obstruction theory, one can show that any principal $\mathbb{R}$-bundle over a base which is a manifold is trivialisable, i.e. has a global section.[18] $\qquad\triangle$

---

[18] The general statement is that any locally trivial (topological) fibre bundle with contractible fibre type and a base that is a CW complex has a global section. In case all involved objects are smooth, there exists a smooth global section.

Proposition 5.39 *Let $\sigma \in \Gamma(M, \tilde{M})$ be a global section. Given a unit timelike vector field v on (M, τ, h), let $\tilde{v}$ be its invariant null lift to $\tilde{M}$, and consider the corresponding principal connection $\tilde{g}(\tilde{v}, \cdot)$. Consider the pullback*

$$a := -\sigma^*(\tilde{g}(\tilde{v}, \cdot)). \tag{5.98}$$

Under a change of vector field v, this one-form transforms as the local representative of a Bargmann form on (M, τ, h). Thus, any choice of global section $\sigma \in \Gamma(M, \tilde{M})$ induces a Bargmann form on M.

Proof Let w be a spacelike vector field on (M, τ, h). We need to show that under the Milne boost $v \mapsto v - w$, the form a transforms according to (5.13a), i.e. as $a \mapsto a + \underset{v}{h}(w, \cdot) + \frac{1}{2}|w|^2\tau$.

Let $\tilde{w}$ be the unique invariant lift of w such that $\tilde{v} - \tilde{w}$ is the unique invariant null lift of $v - w$. Then, by construction of the form a, under the Milne boost $v \mapsto v - w$ it transforms as

$$a = -\sigma^*(\tilde{g}(\tilde{v}, \cdot)) \mapsto -\sigma^*(\tilde{g}(\tilde{v} - \tilde{w}, \cdot)) = a + \sigma^*(\tilde{g}(\tilde{w}, \cdot)). \tag{5.99}$$

This means we have to show that $\sigma^*(\tilde{g}(\tilde{w}, \cdot)) = \underset{v}{h}(w, \cdot) + \frac{1}{2}|w|^2\tau$.

Since w is spacelike, by definition of τ we have

$$\tilde{g}(\tilde{w}, \xi) = (\pi^*\tau)(\tilde{w}) = \tau(w) \circ \pi = 0, \tag{5.100}$$

i.e. the invariant one-form $\tilde{g}(\tilde{w}, \cdot) \in \Omega^1(\tilde{M})$ vanishes on ξ. Therefore, by Lemma 5.31 there is a *unique* one-form $\alpha \in \Omega^1(M)$ satisfying

$$\pi^*\alpha = \tilde{g}(\tilde{w}, \cdot). \tag{5.101}$$

So we are left to show that this α is given by

$$\alpha = \underset{v}{h}(w, \cdot) + \frac{1}{2}|w|^2\tau. \tag{5.102}$$

Since any vector field on M can be decomposed into its spacelike and timelike parts with respect to v, it is enough to show that (5.102) holds when applied to spacelike vector fields on the one hand, and applied to v on the other hand.

For any spacelike vector field X on (M, τ, h) and an invariant lift $\tilde{X}$, by Proposition 5.35 we have $\tilde{g}(\tilde{w}, \tilde{X}) = {}^{(n)}h(w, X) \circ \pi$. Combined with the definition (5.101) of α, this shows that $\alpha(X) = {}^{(n)}h(w, X) = \underset{v}{h}(w, X) = (\underset{v}{h}(w, \cdot) + \frac{1}{2}|w|^2\tau)(X)$, i.e. that (5.102) holds when applied to spacelike vector fields.

Since both $\tilde{v}$ and $\tilde{v} - \tilde{w}$ are null, we have

$$0 = \tilde{g}(\tilde{v} - \tilde{w}, \tilde{v} - \tilde{w}) = \tilde{g}(\tilde{v}, \tilde{v}) - 2\tilde{g}(\tilde{w}, \tilde{v}) + \tilde{g}(\tilde{w}, \tilde{w}) = -2\tilde{g}(\tilde{w}, \tilde{v}) + \tilde{g}(\tilde{w}, \tilde{w}). \tag{5.103}$$

Therefore, we have

$$\tilde{g}(\tilde{w}, \tilde{v}) = \frac{1}{2}\tilde{g}(\tilde{w}, \tilde{w}) = \frac{1}{2}{}^{(n)}h(w, w) \circ \pi = \frac{1}{2}|w|^2 \circ \pi \qquad (5.104)$$

by Proposition 5.35. Combined with the definition (5.101) of α, this shows that $\alpha(v) = \frac{1}{2}|w|^2 = (\underset{v}{h}(w, \cdot) + \frac{1}{2}|w|^2 \tau)(v)$, i.e. that (5.102) holds when applied to v. $\qquad\qquad\square$

Remark 5.40 Note that the Bargmann form from Proposition 5.39 depends on the choice of global section σ. Fixing the unit timelike vector field v, the corresponding form a was defined as the negative of the local representative of the principal connection $\tilde{g}(\tilde{v}, \cdot)$ on $\tilde{M}$ with respect to σ. When changing from σ to a new section σ_2 with $\sigma_2(p) = \Phi_{-\chi(p)}(\sigma(p))$, where $\chi \in C^\infty(M)$ and Φ denotes the flow of ξ, the form therefore changes according to $a = -\sigma^*(\tilde{g}(\tilde{v}, \cdot)) \mapsto -\sigma_2^*(\tilde{g}(\tilde{v}, \cdot)) = a + d\chi$.

This is precisely a type I gauge transformation of the Bargmann form (Construction 5.19). In this sense, type I gauge transformations arise naturally as 'gauge redundancies' of the null reduction framework—they are related to an arbitrary choice that must be made in the passage from $n + 2$ to $n + 1$ dimensions. $\qquad\qquad\triangle$

Construction 5.41 Consider the Levi-Civita connection $\tilde{\nabla}$ of $\tilde{g}$. We want to define a linear connection ∇ on M via invariant lifts, i.e. by demanding

$$\widetilde{\nabla_X Y} = \tilde{\nabla}_{\tilde{X}} \tilde{Y} \qquad (5.105)$$

for vector fields X, Y on M and some choice of invariant lifts $\tilde{X}, \tilde{Y}$. Since $\tilde{\nabla}_{\tilde{X}} \tilde{Y}$ is determined by $\tilde{g}$, $\tilde{X}$ and $\tilde{Y}$ in terms of tensor products, contraction, and (exterior) differentiation of scalar functions, we have $\mathcal{L}_\xi(\tilde{\nabla}_{\tilde{X}} \tilde{Y}) = 0$. Thus, $D\pi|_q((\tilde{\nabla}_{\tilde{X}} \tilde{Y})|_q)$ is independent of the choice of $q \in \pi^{-1}(p)$ for fixed $p \in M$, such that $\tilde{\nabla}_{\tilde{X}} \tilde{Y}$ is indeed an invariant lift of some vector field on M.

However, in order for $\nabla_X Y$ to be well-defined by (5.105), we need the projection $D\pi \circ \tilde{\nabla}_{\tilde{X}} \tilde{Y}$ to M to be independent of the choice of lifts $\tilde{X}, \tilde{Y}$. Considering different lifts $\tilde{X}_2 = \tilde{X} + (\lambda_X \circ \pi) \cdot \xi$, $\tilde{Y}_2 = \tilde{Y} + (\lambda_Y \circ \pi) \cdot \xi$ with real-valued functions λ_X, λ_Y on M, we obtain

$$\begin{aligned}
\tilde{\nabla}_{\tilde{X}_2} \tilde{Y}_2 &= \tilde{\nabla}_{\tilde{X}_2}(\tilde{Y} + (\lambda_Y \circ \pi) \cdot \xi) \\
&= \tilde{\nabla}_{\tilde{X}_2} \tilde{Y} + (\lambda_Y \circ \pi) \cdot \tilde{\nabla}_{\tilde{X}_2} \xi + \tilde{X}_2(\lambda_Y \circ \pi) \cdot \xi \\
&= \tilde{\nabla}_{\tilde{X}} \tilde{Y} + (\lambda_X \circ \pi) \cdot \tilde{\nabla}_\xi \tilde{Y} + (\lambda_Y \circ \pi) \cdot \tilde{\nabla}_{\tilde{X}} \xi + (\lambda_X \lambda_Y \circ \pi) \cdot \tilde{\nabla}_\xi \xi \\
&\quad + (D\pi \circ \tilde{X}_2)(\lambda_Y) \cdot \xi \\
&= \tilde{\nabla}_{\tilde{X}} \tilde{Y} + (\lambda_X \circ \pi) \cdot \tilde{\nabla}_\xi \tilde{Y} + (\lambda_Y \circ \pi) \cdot \tilde{\nabla}_{\tilde{X}} \xi + (\lambda_X \lambda_Y \circ \pi) \cdot \tilde{\nabla}_\xi \xi \\
&\quad + (X(\lambda_Y) \circ \pi) \cdot \xi. \qquad (5.106)
\end{aligned}$$

Therefore, the projections to M agree, i.e. $\mathrm{D}\pi \circ \tilde{\nabla}_{\tilde{X}_2}\tilde{Y}_2 = \mathrm{D}\pi \circ \tilde{\nabla}_{\tilde{X}}\tilde{Y}$, if and only if

$$\lambda_X \cdot (\mathrm{D}\pi \circ \tilde{\nabla}_\xi \tilde{Y}) + \lambda_Y \cdot (\mathrm{D}\pi \circ \tilde{\nabla}_{\tilde{X}}\xi) + \lambda_X \lambda_Y \cdot (\mathrm{D}\pi \circ \nabla_\xi \xi) = 0. \tag{5.107}$$

This shows that the connection ∇ on M is well-defined by (5.105) if and only if (5.107) holds for all real-valued functions λ_X, λ_Y on M and all invariant vector fields $\tilde{X}, \tilde{Y}$ on $\tilde{M}$.

Taking $\lambda_X = 0$, $\lambda_Y = 1$, the latter statement implies in particular that

$$\tilde{\nabla}_{\tilde{X}}\xi \propto \xi \text{ for all invariant vector fields } \tilde{X} \text{ on } \tilde{M}. \tag{5.108}$$

Since any vector $\tilde{w} \in T\tilde{M}$ can be extended to an invariant vector field, this is equivalent to

$$\tilde{\nabla}_{\tilde{w}}\xi \propto \xi \text{ for all vectors } \tilde{w} \in T\tilde{M}. \tag{5.109}$$

In components, this means that

$$\tilde{\nabla}_{\tilde{\mu}}\xi^{\tilde{\nu}} = \eta_{\tilde{\mu}}\xi^{\tilde{\nu}} \tag{5.110}$$

for some one-form $\eta \in \Omega^1(\tilde{M})$, where indices with a tilde denote components of tensor fields on $\tilde{M}$. Now ξ is Killing, $\tilde{\nabla}_{(\tilde{\mu}}\xi_{\tilde{\nu})} = 0$, from which we obtain $\eta_{(\tilde{\mu}}\xi_{\tilde{\nu})} = 0$. Due to ξ vanishing nowhere, this implies $\eta = 0$, i.e. $\tilde{\nabla}\xi = 0$.

Note that conversely $\tilde{\nabla}\xi = 0$ implies that (5.107) holds for all functions λ_X, λ_Y and invariant vector fields $\tilde{X}, \tilde{Y}$: the second and third terms of (5.107) vanish directly, and invariance of $\tilde{Y}$ means $0 = \mathcal{L}_\xi \tilde{Y} = [\xi, \tilde{Y}]$, i.e. (due to $\tilde{\nabla}$ being torsion-free) $\tilde{\nabla}_\xi \tilde{Y} = \tilde{\nabla}_{\tilde{Y}}\xi$, such that the first term of (5.107) vanishes as well.

We have thus proved that ∇ is well-defined by (5.105) if and only if $\tilde{\nabla}\xi = 0$. Such a Lorentzian manifold admitting a covariantly constant null vector field is called a *parallel wave spacetime* (a spacetime describing parallelly propagating waves) [11]. This means we have shown that the null-symmetric Lorentzian manifolds whose quotient Galilei manifold inherits a well-defined connection are precisely those which are parallel waves (with the wave vector field being the symmetry generator).

We still need to show that $\nabla_X Y$ as defined by (5.105) is $C^\infty(M)$-linear in X and satisfies the derivation property in Y, such that it indeed defines a connection. $\mathbb{R}$-linearity follows immediately from that of $\tilde{\nabla}$. Further, choosing invariant lifts $\tilde{X}, \tilde{Y}$ of X, Y, for a function f on M we know that $(f \circ \pi) \cdot \tilde{X}$ is an invariant lift of fX (and analogously for Y). Therefore we obtain

$$\widetilde{\nabla_{fX}Y} = \tilde{\nabla}_{(f\circ\pi)\cdot\tilde{X}}\tilde{Y} = (f \circ \pi) \cdot \tilde{\nabla}_{\tilde{X}}\tilde{Y} = (f \circ \pi) \cdot \widetilde{\nabla_X Y} = \widetilde{f\nabla_X Y}, \tag{5.111a}$$

$$\begin{aligned}
\widetilde{\nabla_X(fY)} &= \tilde{\nabla}_{\tilde{X}}((f \circ \pi) \cdot \tilde{Y}) \\
&= (f \circ \pi) \cdot \tilde{\nabla}_{\tilde{X}}\tilde{Y} + (\tilde{X}(f \circ \pi)) \cdot \tilde{Y} \\
&= (f \circ \pi) \cdot \tilde{\nabla}_{\tilde{X}}\tilde{Y} + ((\mathrm{D}\pi \circ \tilde{X})(f)) \cdot \tilde{Y} \\
&= (f \circ \pi) \cdot \tilde{\nabla}_{\tilde{X}}\tilde{Y} + (X(f) \circ \pi) \cdot \tilde{Y} \\
&= \widetilde{f\nabla_X Y} + \widetilde{X(f) \cdot Y}.
\end{aligned} \tag{5.111b}$$

This shows $\nabla_{fX} Y = f \nabla_X Y$ and $\nabla_X(fY) = f\nabla_X Y + X(f) \cdot Y$ as desired. $\triangle$

Proposition 5.42 *Let $\tilde{\nabla}\xi = 0$, where $\tilde{\nabla}$ is the Levi-Civita connection of $\tilde{g}$, and consider the linear connection ∇ induced on M by $\tilde{\nabla}$.*

(i) ∇ is a Galilei connection on (M, τ, h).
(ii) ∇ is torsion-free.
(iii) The extended torsion of ∇ with respect to any Bargmann form induced by a section of $\tilde{M}$ vanishes. In particular, ∇ is Newtonian.

Proof (i) Compatibility of ∇ with τ and h follows from compatibility of $\tilde{\nabla}$ with $\tilde{g}$ and $\tilde{\nabla}\xi = 0$. Explicitly, this works as follows.

For both the cases of τ and h, we will need to express $\nabla\alpha$, for a one-form $\alpha \in \Omega^1(M)$, in terms of $\tilde{\nabla}$. So let X, Y be vector field on M with invariant lifts $\tilde{X}$, $\tilde{Y}$. We have

$$(\nabla_X \alpha)(Y) = X(\alpha(Y)) - \alpha(\nabla_X Y), \tag{5.112}$$

which implies

$$\begin{aligned}
(\pi^*(\nabla_X\alpha))(\tilde{Y}) &= (\nabla_X\alpha)(Y) \circ \pi \\
&= X(\alpha(Y)) \circ \pi - \alpha(\nabla_X Y) \circ \pi \\
&= \tilde{X}(\alpha(Y) \circ \pi) - (\pi^*\alpha)(\widetilde{\nabla_X Y}) \\
&= \tilde{X}((\pi^*\alpha)(\tilde{Y})) - (\pi^*\alpha)(\tilde{\nabla}_{\tilde{X}}\tilde{Y}) \\
&= (\tilde{\nabla}_{\tilde{X}}(\pi^*\alpha))(\tilde{Y}). \tag{5.113}
\end{aligned}$$

This shows that $\pi^*(\nabla_X\alpha) = \tilde{\nabla}_{\tilde{X}}(\pi^*\alpha)$ (which by Lemma 5.31 determines $\nabla_X\alpha$ uniquely).

Now τ is defined by $\pi^*\tau = \tilde{g}(\xi, \cdot)$. Therefore, for any vector field X on M we have $\pi^*(\nabla_X\tau) = \tilde{\nabla}_{\tilde{X}}(\pi^*\tau) - \tilde{\nabla}_{\tilde{X}}(\tilde{g}(\xi, \cdot)) = 0$ due to $\tilde{\nabla}\tilde{g} = 0$ and $\tilde{\nabla}\xi = 0$. This shows $\nabla\tau = 0$.

Further, h is defined by $\tilde{g}^{-1}(\pi^*\alpha, \pi^*\beta) = h(\alpha, \beta) \circ \pi$ for any two one-forms $\alpha, \beta \in \Omega^1(M)$. Thus, for any vector field X on M we obtain

$$\begin{aligned}
X(h(\alpha, \beta)) \circ \pi &= \tilde{X}(h(\alpha, \beta) \circ \pi) \\
&= \tilde{X}(\tilde{g}^{-1}(\pi^*\alpha, \pi^*\beta)) \\
&= \tilde{g}^{-1}(\tilde{\nabla}_{\tilde{X}}(\pi^*\alpha), \pi^*\beta) + \tilde{g}^{-1}(\pi^*\alpha, \tilde{\nabla}_{\tilde{X}}(\pi^*\beta)) \\
&= \tilde{g}^{-1}(\pi^*(\nabla_X\alpha), \pi^*\beta) + \tilde{g}^{-1}(\pi^*\alpha, \pi^*(\nabla_X\beta)) \\
&= (h(\nabla_X\alpha, \beta) + h(\alpha, \nabla_X\beta)) \circ \pi, \tag{5.114}
\end{aligned}$$

where we have used $\tilde{\nabla}\tilde{g}^{-1} = 0$. This shows $\nabla h = 0$.

(ii) For any two vector fields X, Y on M, by Remark 5.34 we have $[\tilde{X}, \tilde{Y}] = \widetilde{[X, Y]}$. Thus, using that $\tilde{\nabla}$ is torsion-free, we obtain

$$\widetilde{[X, Y]} = [\tilde{X}, \tilde{Y}] = \tilde{\nabla}_{\tilde{X}}\tilde{Y} - \tilde{\nabla}_{\tilde{Y}}\tilde{X} = \widetilde{\nabla_X Y} - \widetilde{\nabla_Y X}. \tag{5.115}$$

This shows $[X, Y] = \nabla_X Y - \nabla_Y X$, i.e. torsion-freeness of ∇.

(iii) Let $\sigma \in \Gamma(M, \tilde{M})$ be a global section, and consider the corresponding induced Bargmann form. We have to show that the mass torsion of ∇ with respect to this Bargmann form vanishes. Let v be a unit timelike vector field on M. The representative of the Bargmann form with respect to v is $a = -\sigma^*\omega$, where $\omega = \tilde{g}(\tilde{v}, \cdot)$ and $\tilde{v}$ is the invariant null lift of v to $\tilde{M}$. The local representative of the mass torsion is $\mathrm{d}a + \Omega$, where Ω is the Newton–Coriolis form of ∇ with respect to v. We thus have to show that Ω and $-\mathrm{d}a$ are equal.

To compute $-\mathrm{d}a$, we will first prove a useful identity. Let $w \in T_p M$ be any vector, and $\tilde{w} \in T_q \tilde{M}$ with $\pi(q) = p$ a lift of w, i.e. a vector satisfying $\mathrm{D}\pi|_q(\tilde{w}) = w$. By construction of ∇, the vector $\tilde{\nabla}_{\tilde{w}}\tilde{v} \in T_q M$ is a lift of $\nabla_w v \in T_p M$. Since ∇ is compatible with τ, we know that $\nabla_w v$ is spacelike. Thus, if $u \in \ker \tau|_p$ is another spacelike vector with a lift $\tilde{u} \in T_q \tilde{M}$, $\mathrm{D}\pi|_q(\tilde{u}) = u$, by Proposition 5.35 (applied pointwise) we have

$$\tilde{g}(\tilde{\nabla}_{\tilde{w}}\tilde{v}, \tilde{u}) = {}^{(n)}h(\nabla_w v, u). \tag{5.116}$$

Now we turn to the computation of $-\mathrm{d}a$. By definition, we have

$$-\mathrm{d}a = \mathrm{d}(\sigma^*\omega) = \sigma^*(\mathrm{d}\omega). \tag{5.117}$$

Since $\tilde{\nabla}$ is torsion-free, we may express $\mathrm{d}\omega$ as twice the anti-symmetrised covariant derivative $\nabla\omega$, i.e. for arbitrary vectors $\tilde{w}, \tilde{u} \in T_q \tilde{M}$ we have

$$\mathrm{d}\omega(\tilde{w}, \tilde{u}) = (\tilde{\nabla}_{\tilde{w}}\omega)(\tilde{u}) - (\tilde{\nabla}_{\tilde{u}}\omega)(\tilde{w}). \tag{5.118}$$

Inserting the concrete form of ω, and using $\tilde{\nabla}\tilde{g} = 0$, this yields

$$\mathrm{d}\omega(\tilde{w}, \tilde{u}) = \tilde{g}(\tilde{\nabla}_{\tilde{w}}\tilde{v}, \tilde{u}) - \tilde{g}(\tilde{\nabla}_{\tilde{u}}\tilde{v}, \tilde{w}). \tag{5.119}$$

For $w, u \in \ker \tau|_p$ spacelike vectors, and $\tilde{w}, \tilde{u} \in T_q \tilde{M}$ arbitrary lifts thereof, using (5.116) this implies

$$\begin{aligned}
\mathrm{d}\omega(\tilde{w}, \tilde{u}) &= \tilde{g}(\tilde{\nabla}_{\tilde{w}}\tilde{v}, \tilde{u}) - \tilde{g}(\tilde{\nabla}_{\tilde{u}}\tilde{v}, \tilde{u}) \\
&= {}^{(n)}h(\nabla_w v, u) - {}^{(n)}h(\nabla_u v, w). \tag{5.120a}
\end{aligned}$$

Further, for $\hat{v} = \tilde{v}|_q + \lambda \xi|_q$ a general vector in $T_q \tilde{M}$ projecting to v (that is not necessarily null), again using (5.116) we obtain

$$
\begin{aligned}
\mathrm{d}\omega(\hat{v}, \tilde{w}) &= \tilde{g}(\tilde{\nabla}_{\hat{v}}\tilde{v}, \tilde{w}) - \tilde{g}(\tilde{\nabla}_{\tilde{w}}\tilde{v}, \hat{v}) \\
&= {}^{(n)}h(\nabla_v v, w) - \tilde{g}(\tilde{\nabla}_{\tilde{w}}\tilde{v}, \tilde{v}|_q) - \lambda \tilde{g}(\tilde{\nabla}_{\tilde{w}}\tilde{v}, \xi|_q) \\
&= {}^{(n)}h(\nabla_v v, w) - \frac{1}{2}\tilde{w}(\underbrace{\tilde{g}(\tilde{v}, \tilde{v})}_{=0}) - \lambda \underbrace{(\pi^*\tau)(\tilde{\nabla}_{\tilde{w}}\tilde{v})}_{=\tau(\nabla_w v)=0} \\
&= {}^{(n)}h(\nabla_v v, w).
\end{aligned}
\tag{5.120b}
$$

Since $\mathrm{D}\sigma(v|_p), \mathrm{D}\sigma(w), \mathrm{D}\sigma(u) \in T_{\sigma(p)}\tilde{M}$ project to $v|_p, w, u$, respectively, this shows that $-\mathrm{d}a|_p = \sigma^*(\mathrm{d}\omega)|_p$ is given by

$$
-\mathrm{d}a|_p(w, u) = {}^{(n)}h(\nabla_w v, u) - {}^{(n)}h(\nabla_u v, w)
\tag{5.121a}
$$

on spacelike vectors, and by

$$
-\mathrm{d}a|_p(v|_p, w) = {}^{(n)}h(\nabla_v v, w)
\tag{5.121b}
$$

on $v|_p$ and one spacelike vector.

This is precisely the behaviour of the Newton–Coriolis form Ω of ∇ with respect to v when evaluated on either a pair of spacelike vectors or on v and one spacelike vector. This shows that $-\mathrm{d}a = \Omega$, finishing the proof. $\qquad\square$

Construction 5.43 Assume that the Levi-Civita connection $\tilde{\nabla}$ of $\tilde{g}$ satisfies $\tilde{\nabla}\xi = 0$, and consider the induced linear connection ∇ on M. Let $\tilde{R}$ be the curvature tensor of $\tilde{\nabla}$, and R that of ∇. For any vector fields X, Y, Z on M and invariant lifts $\tilde{X}, \tilde{Y}, \tilde{Z}$ of these to $\tilde{M}$, by definition of ∇ and by Remark 5.34 we have

$$
\begin{aligned}
\tilde{R}(\tilde{X}, \tilde{Y})\tilde{Z} &= \tilde{\nabla}_{\tilde{X}}\tilde{\nabla}_{\tilde{Y}}\tilde{Z} - \tilde{\nabla}_{\tilde{Y}}\tilde{\nabla}_{\tilde{X}}\tilde{Z} - \tilde{\nabla}_{[\tilde{X},\tilde{Y}]}\tilde{Z} \\
&= \tilde{\nabla}_{\tilde{X}}\widetilde{\nabla_Y Z} - \tilde{\nabla}_{\tilde{Y}}\widetilde{\nabla_X Z} - \tilde{\nabla}_{\widetilde{[X,Y]}}\tilde{Z} \\
&= \widetilde{\nabla_X \nabla_Y Z} - \widetilde{\nabla_Y \nabla_X Z} - \widetilde{\nabla_{[X,Y]}Z} \\
&= \widetilde{R(X, Y)Z}.
\end{aligned}
\tag{5.122a}
$$

Formulated in words, applying the curvature tensor of $\tilde{\nabla}$ to invariant lifts of the original vector fields gives an invariant lift of the curvature tensor of ∇ applied to the original vector fields. Furthermore, due to $\tilde{\nabla}\xi = 0$, the curvature satisfies

$$
\tilde{R}(\cdot, \cdot)\xi = 0.
\tag{5.122b}
$$

Using the symmetries of $\tilde{R}$, this implies that *all* contractions of $\tilde{R}$ with ξ vanish, i.e.

$$\tilde{R}(\xi, \cdot)\cdot = -\tilde{R}(\cdot, \xi)\cdot = 0, \tag{5.122c}$$

$$\tilde{g}(\xi, \tilde{R}(\cdot, \cdot)\cdot) = 0. \tag{5.122d}$$

We may use this to prove that ∇ is Newtonian in a more direct way. First, let α be a one-form on M. Since both $\tilde{g}$ and $\pi^*\alpha$ are invariant under the flow of ξ, the vector field $\tilde{g}^{-1}(\pi^*\alpha, \cdot)$ on $\tilde{M}$ is invariant under the flow as well. For any other one-form β on M, by definition of h we have $\beta(\mathrm{D}\pi|_q(\tilde{g}^{-1}|_q(\pi^*\alpha|_q, \cdot))) = \tilde{g}^{-1}|_q(\pi^*\alpha|_q, \pi^*\beta|_q) = h|_p(\alpha|_p, \beta|_p)$ for $q \in \tilde{M}$, $p = \pi(q)$. This shows that $\tilde{g}^{-1}(\pi^*\alpha, \cdot)$ is an invariant lift of $h(\alpha, \cdot)$, i.e. symbolically

$$\tilde{g}^{-1}(\pi^*\alpha, \cdot) = \widetilde{h(\alpha, \cdot)}. \tag{5.123}$$

Now consider the symmetry in pairs $\tilde{R}^{\tilde{\mu}}{}_{\tilde{\rho}}{}^{\tilde{\nu}}{}_{\tilde{\sigma}} = \tilde{R}^{\tilde{\nu}}{}_{\tilde{\sigma}}{}^{\tilde{\mu}}{}_{\tilde{\rho}}$ of $\tilde{R}$, where the third index was raised with $\tilde{g}^{\tilde{\mu}\tilde{\nu}}$. Formulated in a component-free way, this reads

$$\tilde{\beta}(\tilde{R}(\tilde{g}^{-1}(\tilde{\alpha}, \cdot), \tilde{Y})\tilde{Z}) = \tilde{\alpha}(\tilde{R}(\tilde{g}^{-1}(\tilde{\beta}, \cdot), \tilde{Z})\tilde{Y}) \tag{5.124}$$

for any vector fields $\tilde{Y}$, $\tilde{Z}$ and one-forms $\tilde{\alpha}$, $\tilde{\beta}$ on $\tilde{M}$. Applying this to the case that $\tilde{Y}$, $\tilde{Z}$ are invariant lifts of vector fields Y, Z on M and that $\tilde{\alpha} = \pi^*\alpha$, $\tilde{\beta} = \pi^*\beta$ for one-forms α, β on M, using (5.123) we obtain

$$\pi^*\beta\left(\tilde{R}\left(\widetilde{h(\alpha, \cdot)}, \tilde{Y}\right)\tilde{Z}\right) = \pi^*\alpha\left(\tilde{R}\left(\widetilde{h(\beta, \cdot)}, \tilde{Z}\right)\tilde{Y}\right). \tag{5.125}$$

Using the relationship (5.122a) between the curvature tensors, this implies

$$\beta(R(h(\alpha, \cdot), Y)Z) = \alpha(R(h(\beta, \cdot), Z)Y). \tag{5.126}$$

This is symmetry in pairs $R^{\mu}{}_{\rho}{}^{\nu}{}_{\sigma} = R^{\nu}{}_{\sigma}{}^{\mu}{}_{\rho}$ of R, where the third index was raised with $h^{\mu\nu}$—i.e. precisely the condition that ∇ be Newtonian.

Further, we can use the expressions (5.122) for the curvature tensor of $\tilde{\nabla}$ to relate its Ricci tensor $\widetilde{\mathrm{Ric}}$ to the Ricci tensor Ric of ∇: we are going to show that they satisfy

$$\widetilde{\mathrm{Ric}} = \pi^*\mathrm{Ric}, \tag{5.127a}$$

i.e. that for any $\tilde{w}, \tilde{u} \in T_q M$ we have

$$\widetilde{\mathrm{Ric}}|_q(\tilde{w}, \tilde{u}) = \mathrm{Ric}|_{\pi(q)}(\mathrm{D}\pi|_q(\tilde{w}), \mathrm{D}\pi|_q(\tilde{u})). \tag{5.127b}$$

This may be seen as follows. Let $\tilde{w}, \tilde{u} \in T_q \tilde{M}$, and denote their projections to M by $w = \mathrm{D}\pi|_q(\tilde{w})$, $u = \mathrm{D}\pi|_q(\tilde{u}) \in T_p M$ (where $p = \pi(q)$). By definition, we have

$$\widetilde{\mathrm{Ric}}|_q(\tilde{w}, \tilde{u}) = \mathrm{tr}_{T_q\tilde{M}}(\tilde{R}|_q(\cdot, \tilde{u})\tilde{w}), \quad \mathrm{Ric}|_p(w, u) = \mathrm{tr}_{T_pM}(R|_p(\cdot, u)w). \tag{5.128}$$

We need to show that these two expressions are equal. To see this, consider the linear map $f : T_p M \to T_q \tilde{M}$ defined by

$$f(v) := \tilde{R}|_q(\tilde{v}, \tilde{u})\tilde{w}, \quad \text{where } \tilde{v} \in T_q \tilde{M} \text{ satisfies } \mathrm{D}\pi|_q(\tilde{v}) = v. \tag{5.129}$$

Since $\tilde{v}$ is unique up to a multiple of ξ, on which R vanishes, f is well-defined. Now by definition, for any $\tilde{v} \in T_q \tilde{M}$ we have $f(\mathrm{D}\pi|_q(\tilde{v})) = \tilde{R}|_q(\tilde{v}, \tilde{u})\tilde{w}$, which shows

$$f \circ \mathrm{D}\pi|_q = \tilde{R}|_q(\cdot, \tilde{u})\tilde{w}. \tag{5.130a}$$

Further, evaluating (5.122a) pointwise shows that for any $v \in T_p M$ we have $\mathrm{D}\pi|_q(f(v)) = \mathrm{D}\pi|_q(\tilde{R}|_q(\tilde{v}, \tilde{u})\tilde{w}) = R|_p(v, u)w$. This means that

$$\mathrm{D}\pi|_q \circ f = R|_p(\cdot, u)w. \tag{5.130b}$$

By the cyclic property of the trace, we have

$$\mathrm{tr}_{T_q \tilde{M}}(f \circ \mathrm{D}\pi|_q) = \mathrm{tr}_{T_p M}(\mathrm{D}\pi|_q \circ f). \tag{5.131}$$

This shows that the two expressions in (5.128) are equal, finishing the proof of (5.127).

Finally, we may now use (5.127) to relate the Newton–Cartan field equation for (M, τ, h, ∇) to the Einstein equation for $(\tilde{M}, \tilde{g})$. The Newton–Cartan field equation reads

$$\mathrm{Ric} = 4\pi G \rho \tau \otimes \tau. \tag{5.132}$$

According to (5.127), this holds if and only if the Ricci tensor of $\tilde{\nabla}$ is given by

$$\widetilde{\mathrm{Ric}} = 4\pi G(\rho \circ \pi)\pi^*\tau \otimes \pi^*\tau = 4\pi G(\rho \circ \pi)\tilde{g}(\xi, \cdot) \otimes \tilde{g}(\xi, \cdot), \tag{5.133a}$$

or in components

$$\tilde{R}_{\mu\nu} = 4\pi G(\rho \circ \pi)\xi_{\tilde{\mu}}\xi_{\tilde{\nu}} \tag{5.133b}$$

Since ξ is null, this is equivalent to the Einstein tensor satisfying

$$\tilde{R}_{\tilde{\mu}\tilde{\nu}} - \frac{1}{2}\tilde{R}\tilde{g}_{\tilde{\mu}\tilde{\nu}} = 4\pi G(\rho \circ \pi)\xi_{\tilde{\mu}}\xi_{\tilde{\nu}}. \tag{5.134}$$

Hence, the Newton–Cartan field equation for (M, τ, h, ∇) is equivalent to the Einstein equation

$$\tilde{R}_{\tilde{\mu}\tilde{\nu}} - \frac{1}{2}\tilde{R}\tilde{g}_{\tilde{\mu}\tilde{\nu}} = \kappa \tilde{T}_{\tilde{\mu}\tilde{\nu}} \tag{5.135}$$

for $(\tilde{M}, \tilde{g})$ if and only if the energy–momentum tensor is given by

$$\tilde{T}_{\tilde{\mu}\tilde{\nu}} = \frac{4\pi G}{\kappa}(\rho \circ \pi)\xi_{\tilde{\mu}}\xi_{\tilde{\nu}}, \tag{5.136}$$

i.e. if and only if the matter in our higher-dimensional parallel wave spacetime is *null dust* in the direction of ξ.   △

Construction 5.44 Conversely to the construction described in this section, we may also start with a Galilei manifold (M, τ, h), $\dim M = n + 1$, with a Bargmann form $\boldsymbol{a}$, and from this construct a null-symmetric Lorentzian manifold $(\tilde{M}, \tilde{g}, \xi)$ from which (M, τ, h) and $\boldsymbol{a}$ arise by null reduction. This works as follows.

We define $\tilde{M} = M \times \mathbb{R}$, and denote the coordinate function that projects on the $\mathbb{R}$ factor by $s \colon \tilde{M} \to \mathbb{R}$. The projection onto the M factor we denote by $\pi \colon \tilde{M} \to M$.

On $\tilde{M}$, let ξ be the vector field pointing along the $\mathbb{R}$ direction, i.e. the field whose flow lines are the curves $\mathbb{R} \ni s \mapsto (p, s) \in \tilde{M}$ of constant $p \in M$ and varying s. Put differently, if (x^μ) are (local) coordinates for M, then $(x^\mu \circ \pi, s)$ are (local) coordinates for $\tilde{M}$, and in any such coordinates ξ is given by $\xi = \frac{\partial}{\partial s}$.

Let $\hat{v}$ be the boost-invariant unit timelike vector field on (M, τ, h) determined by the Bargmann form $\boldsymbol{a}$ according to Construction 5.15, and $\hat{\phi}$ the corresponding function on M determined by $\hat{a} = \hat{\phi}\tau$ (where $\hat{a}$ is the local representative of $\boldsymbol{a}$ with respect to $\hat{v}$).

On $\tilde{M}$, we define a symmetric covariant 2-tensor field $\tilde{g} \in \Gamma(\bigvee^2 T^*\tilde{M})$ as

$$\tilde{g} := \pi^*\left(\underset{\hat{v}}{h} - 2\hat{\phi}\tau \otimes \tau \right) + \pi^*\tau \otimes \mathrm{d}s + \mathrm{d}s \otimes \pi^*\tau. \tag{5.137}$$

By definition, we have $\tilde{g}(\xi, \xi) = 0$ and $\mathcal{L}_\xi \tilde{g} = 0$. However, we do not yet know that $\tilde{g}$ is a Lorentzian metric.

Using the decomposition $T_{(p,s)}\tilde{M} = T_p M \oplus T_s \mathbb{R}$, we define the vector field $\hat{V} \in \Gamma(T\tilde{M})$ by $\hat{V}|_{(p,s)} = (\hat{v}|_p, 0)$ (which is an invariant lift of $\hat{v}$). Further, decomposing also the cotangent spaces as $T^*_{(p,s)}\tilde{M} = T^*_p M \oplus T^*_s \mathbb{R}$, we define $H \in \Gamma(\bigvee^2 T\tilde{M})$ by

$$H|_{(p,s)}((\alpha, \lambda), (\beta, \mu)) := h|_p(\alpha, \beta) \tag{5.138}$$

for $(\alpha, \lambda), (\beta, \mu) \in T^*_{(p,s)}\tilde{M}$. Using these, we define the symmetric contravariant 2-tensor field $\tilde{g}^{-1} \in \Gamma(\bigvee^2 T\tilde{M})$ as

$$\tilde{g}^{-1} := H + \hat{V} \otimes \xi + \xi \otimes \hat{V} + 2(\hat{\phi} \circ \pi) \cdot \xi \otimes \xi. \tag{5.139}$$

By a direct computation, one may check that this is indeed the inverse of $\tilde{g}$, showing that both are non-degenerate.

At each point $(p, s) \in \tilde{M}$, the covector $\pi^*\tau|_{(p,s)} = (\tau|_p, 0) \in T^*_{(p,s)}\tilde{M} = T^*_p M \oplus T^*_s \mathbb{R}$ is null with respect to $\tilde{g}^{-1}|_{(p,s)}$. Its orthogonal complement with respect to $\tilde{g}^{-1}|_{(p,s)}$ is $T^*_p M \oplus \{0\} \subset T^*_{(p,s)}\tilde{M}$, restricted to which $\tilde{g}^{-1}|_{(p,s)}$ is given by $H|_{(p,s)}$. This is positive semidefinite of rank n, with degenerate direction spanned by $\pi^*\tau|_{(p,s)}$. But in general, given any non-degenerate symmetric bilinear form on any vector space, if this form has a null direction, and its restriction to the orthogonal complement of the null direction is positive semidefinite with precisely one degenerate direction (namely the null direction), then the form has Lorentzian signature. Thus, we have shown that $\tilde{g}^{-1}$ has Lorentzian signature; i.e. $\tilde{g}$ is a Lorentzian metric on $\tilde{M}$.

We have thus indeed constructed a null-symmetric Lorentzian manifold $(\tilde{M}, \tilde{g}, \xi)$. By construction, τ and h satisfy the equations $\pi^*\tau = \tilde{g}(\xi, \cdot)$ and $\tilde{g}^{-1}(\pi^*\alpha, \pi^*\beta) = H(\pi^*\alpha, \pi^*\beta) = h(\alpha, \beta) \circ \pi$ for all $\alpha, \beta \in \Omega^1(M)$. Therefore, we indeed recover τ and h when performing null reduction of $(\tilde{M}, \tilde{g}, \xi)$ according to Proposition 5.32.

Finally, the inclusion $\sigma : M \to \tilde{M}, \sigma(p) = (p, 0)$ is a section of π. By Proposition 5.39, this induces a Bargmann form on (M, τ, h). This is the original Bargmann form a we started with: using the definition of $\tilde{g}$, we see that the invariant null lift $\tilde{v}$ of $\hat{v}$ is given by

$$\tilde{v} = \hat{V} + (\hat{\phi} \circ \pi) \cdot \xi. \tag{5.140}$$

Therefore, the representative of the Bargmann form induced by σ with respect to $\hat{v}$ is

$$
\begin{aligned}
-\sigma^*(\tilde{g}(\tilde{v}, \cdot)) &= -\sigma^*\big(-2(\hat{\phi} \circ \pi) \cdot \pi^*\tau + \mathrm{d}s + (\hat{\phi} \circ \pi) \cdot \pi^*\tau\big) \\
&= \hat{\phi} \cdot \sigma^*\pi^*\tau - \sigma^*\mathrm{d}s \\
&= \hat{\phi}\tau - \mathrm{d}(\sigma^*s) \\
&= \hat{\phi}\tau.
\end{aligned}
\tag{5.141}
$$

This is indeed $\hat{a}$, such that the induced Bargmann form is the original one. $\triangle$

References

1. Andringa, R., Bergshoeff, E., Panda, S., de Roo, M.: Newtonian gravity and the Bargmann algebra. Class. Quantum Gravity **28**(10), 105011 (2011). https://doi.org/10.1088/0264-9381/28/10/105011
2. von Blanckenburg, A.L., Schwartz, P.K.: On gauge transformations in twistless-torsional Newton–Cartan geometry. Class Quantum Gravity **42**(22), 225021 (2025). https://doi.org/10.1088/1361-6382/ae1801
3. Duval, C., Burdet, G., Künzle, H.P., Perrin, M.: Bargmann structures and Newton Cartan theory. Phys. Rev. D **31**(8), 1841–1853 (1985). https://doi.org/10.1103/PhysRevD.31.1841
4. Duval, C., Künzle, H.P.: Sur les connexions newtoniennes et l'extension non triviale du groupe de Galilée. C. R. Acad. Sci. A **285**, 813–816 (1977). https://www.researchgate.net/publication/267065597_Sur_les_connexions_newtoniennes_et_l%27extension_non_triviale_du_groupe_de_Galilee
5. Duval, C., Künzle, H.P.: Dynamics of continua and particles from general covariance of Newtonian gravitation theory. Rep. Math. Phys. **13**(3), 351–368 (1978). https://doi.org/10.1016/0034-4877(78)90063-0
6. Duval, C., Künzle, H.P.: Minimal gravitational coupling in the Newtonian theory and the covariant Schrödinger equation. Gen. Relativ. Gravit. **16**(4), 333–347 (1984). https://doi.org/10.1007/BF00762191
7. Geracie, M., Prabhu, K., Roberts, M.M.: Curved non-relativistic spacetimes, Newtonian gravitation and massive matter. J. Math. Phys. **56**(10), 103505 (2015). https://doi.org/10.1063/1.4932967
8. Hansen, D., Hartong, J., Obers, N.A.: Action principle for Newtonian gravity. Phys. Rev. Lett. **122**(6), 061106 (2019). https://doi.org/10.1103/PhysRevLett.122.061106
9. Hansen, D., Hartong, J., Obers, N.A.: Non-relativistic gravity and its coupling to matter. J. High Energy Phys. **2020**(6), 145 (2020). https://doi.org/10.1007/JHEP06(2020)145

10. Julia, B., Nicolai, H.: Null-Killing vector dimensional reduction and Galilean geometrodynamics. Nucl. Phys. B **439**(1), 291–323 (1995). https://doi.org/10.1016/0550-3213(94)00584-2
11. Roche, C., Aazami, A.B., Cederbaum, C.: Exact parallel waves in general relativity. Gen. Relativ. Gravit. **55**(2), 40 (2023). https://doi.org/10.1007/s10714-023-03083-x
12. Schwartz, P.K.: Teleparallel Newton–Cartan gravity. Class. Quantum Gravity **40**(10), 105008 (2023). https://doi.org/10.1088/1361-6382/accc02

Outlook

6

Abstract

In this chapter, we will rather briefly discuss two modern generalisations of 'classical' Newton–Cartan gravity: torsional Newton–Cartan gravity on the one hand, and string Newton–Cartan geometry on the other hand. Since these are very broad fields of active research, by its very nature our brief overview can only provide a crude sketch; for further details, we refer to the original literature.

6.1 Torsional Newton–Cartan Gravity

In so-called *torsional Newton–Cartan gravity/geometry* (TNC gravity/geometry), one considers Galilei manifolds with non-absolute time, i.e. $d\tau \neq 0$. Galilei connections on such manifolds will have non-vanishing timelike torsion—hence the name. Of course, also in the case of absolute time one may consider, according to the classification theorem for Galilei connections (Theorem 2.29), Galilei connections with non-vanishing *spacelike* torsion. This, however, is not commonly called 'torsional Newton–Cartan gravity'; the latter is the established name specifically for situations with non-closed clock form, and hence necessarily non-vanishing timelike torsion.

The most important special case of TNC geometry is so-called *twistless-torsional Newton–Cartan geometry* (TTNC geometry): here one assumes $\tau \wedge d\tau = 0$, such that spatial leaves integrating the spacelike distribution $\ker \tau$ still exist and we have an absolute notion of simultaneity, even though there is no absolute notion of time (if $d\tau \neq 0$). This may be taken as modelling 'gravitational time dilation' in a context in which causality is still Newtonian (i.e. spacetime geometry is still described by Galilei geometry).

Most of our discussion of Galilei geometry in Chaps. 2, 4, 5 did not assume $d\tau = 0$, and the results explicitly take torsion of Galilei connections into account. Therefore, they are directly applicable to the TNC case—or, put differently, we actually dealt with this case most of the time. For example, in our discussion of Bargmann forms, we saw that type I and type II gauge transformations differ in case $d\tau \neq 0$; with

© The Author(s), under exclusive license to Springer Nature Switzerland AG 2026   131
P. K. Schwartz, *Newton–Cartan Gravity*, Lecture Notes in Physics 1044,
https://doi.org/10.1007/978-3-032-03967-5_6

type I being the natural 'geometric redundancies' when considering null reduction of Lorentzian geometry, and type I when considering c^{-1} expansions.

TNC geometry was first introduced in the context of holography by Christensen *et al.* in 2013 (published in 2014) [8, 9]. It has been studied from a more mathematical point of view [2, 13], as well as widely considered in applications of Galilei geometry in condensed matter theory [14, 15, 29] and in aspects of quantum gravity such as holography and null reduction or $c \to \infty$ limits of string theory [5, 6, 12, 22, 25, 26]. Most recently, it has found applications in the coordinate-free description of the post-Newtonian expansion of GR [7, 10, 18–20, 27]. It was this latter context in which type II gauge transformations were discovered.

The two aspects of TNC gravity we will explore here are the following. On the one hand, we discuss the construction of natural Galilei connections on a Galilei manifold with non-absolute time. On the other hand, we will get a glimpse of the $c \to \infty$ limit of Lorentzian spacetimes with respect to non-closed clock forms, and of how this allows to describe 'strong gravity' effects in a theory with Newtonian causality.

Construction 6.1 Let (M, τ, h) be a Galilei manifold and $\boldsymbol{a}$ a Bargmann form on it. By Theorem 5.10, Galilei connections on (M, τ, h) are classified by their extended torsion with respect to $\boldsymbol{a}$. In the case of absolute time, $\mathrm{d}\tau = 0$, there is therefore a natural choice of Galilei connection induced by the choice of Bargmann form, namely the extended-torsion-free one.

However, assuming that $\mathrm{d}\tau \neq 0$, the torsion of any Galilei connection is necessarily non-vanishing. Given a choice of torsion tensor $T \in \Gamma(TM \otimes \bigwedge^2 T^*M)$ satisfying $\tau(T(\cdot, \cdot)) = \mathrm{d}\tau$, we want to find a 'natural' choice of extended torsion, giving rise to a 'natural' Galilei connection (generalising the extended-torsion-free one in the $\mathrm{d}\tau = 0$ case). Put differently, we have to find a 'natural' choice of mass torsion $\boldsymbol{f} \in \Omega^2(G(M))$ that together with the tensorial form $\boldsymbol{T} \in \Omega^2_{\mathrm{Gal}}(G(M), \mathbb{R}^{n+1})$ corresponding to T combines into a $\dot{\rho}$-tensorial form $(\boldsymbol{T}, \boldsymbol{f}) \in \Omega^2_{\dot{\rho}}(G(M), \mathbb{R}^{n+1} \oplus \mathbb{R})$.

Describing $\boldsymbol{f}$ in terms of its local representatives with respect to unit timelike vector fields (as introduced in Remark 5.8), this means the following: we have to find a 'natural' way of associating to each choice of unit timelike vector field v a local two-form f which under a Milne boost $v \mapsto v - w$ transforms as

$$f \mapsto f + \underset{v}{h}(w, T(\cdot, \cdot)) + \frac{1}{2}|w|^2 \mathrm{d}\tau. \tag{6.1}$$

Here we have put the word 'natural' in quotation marks since, as we will see in the following, there are actually *several* natural choices for $\boldsymbol{f}$ (all of which reduce to vanishing extended torsion in the $\mathrm{d}\tau = 0$ case).

(i) From Remark 5.8, we know that the local representatives of the Bargmann form $\boldsymbol{a}$ transform under Milne boosts according to

$$a \mapsto a + \underset{v}{h}(w, \cdot) + \frac{1}{2}|w|^2 \tau. \tag{6.2}$$

Comparing this with the desired transformation behaviour (6.1) for f, we directly see that we may satisfy this by setting

$$f = a(T(\cdot, \cdot)). \tag{6.3}$$

The Galilei connection characterised by this choice of mass torsion (for a given torsion tensor T and Bargmann form a) was dubbed the 'torsional Newtonian connection' by Bekaert and Morand in the article [2].[1]

(ii) We know that the Bargmann form a determines a unit timelike vector field $\hat{v}$ on (M, τ, h) according to Construction 5.15 (the boost-invariant field). We can then make an *arbitrary* choice for $\hat{f}$ (the representative of the mass torsion with respect to $\hat{v}$), thus determining the mass torsion f: for a general unit timelike vector field v, the boost-invariant field $\hat{v}$ is given by $\hat{v} = v - \hat{w}$ with $\hat{w} = -h(a, \cdot)$, such that according to (6.1) we have

$$\hat{f} = f + \underset{v}{h}(\hat{w}, T(\cdot, \cdot)) + \frac{1}{2}|\hat{w}|^2 \mathrm{d}\tau$$
$$= f - a(P(T(\cdot, \cdot))) + \frac{1}{2}h(a, a)\mathrm{d}\tau, \tag{6.4}$$

where $P = \mathrm{id} - v \otimes \tau$ is the spatial projector along v. Conversely, f is determined by $\hat{f}$ as

$$f = \hat{f} + a(P(T(\cdot, \cdot))) - \frac{1}{2}h(a, a)\mathrm{d}\tau$$
$$= \hat{f} + a(T(\cdot, \cdot)) - \hat{\phi}\,\mathrm{d}\tau, \tag{6.5}$$

where $\hat{\phi}$ is the function on M determined by $\hat{a} = \hat{\phi}\tau$ (where $\hat{a}$ is the local representative of a with respect to $\hat{v}$).

From this point of view, i.e. when specifying the mass torsion by choosing $\hat{f}$, the most natural choice is arguably $\hat{f} = 0$. For the mass torsion's local representative with respect to a general unit timelike vector field, this choice yields $f = a(T(\cdot, \cdot)) - \hat{\phi}\,\mathrm{d}\tau$.

This is different from the mass torsion of point (i), which corresponds to $\hat{f} = \hat{\phi}\,\mathrm{d}\tau$.

Now suppose that we are only given the Galilei manifold (M, τ, h) and Bargmann form a, and have to decide which torsion T to choose for our connection (before deciding on the mass torsion). A possible natural choice is $T = \hat{v} \otimes \mathrm{d}\tau$, i.e. taking the spacelike part of T with respect to $\hat{v}$ to vanish. For this specific choice of

---

[1] In reference [2], the notion of mass torsion is not used. However, the Newton–Coriolis form of the connection considered here is $\Omega_{\mu\nu} = f_{\mu\nu} - (\mathrm{d}a)_{\mu\nu} = a_\rho T^\rho{}_{\mu\nu} - 2\partial_{[\mu}a_{\nu]} = 2a_\rho \Gamma^\rho{}_{[\mu}\nu] - 2\partial_{[\mu}a_{\nu]} = -2\nabla_{[\mu}a_{\nu]}$, matching the definition of the 'torsional Newtonian connection' from ref. [2].

torsion, the Galilei connection considered above in point (i) has local mass torsion representative $f = a(\hat{v})\mathrm{d}\tau = (a(v) + h(a, a))\mathrm{d}\tau = (\hat{\phi} + \frac{1}{2}h(a, a))\mathrm{d}\tau$. The $\hat{f} = 0$ connection from point (ii) has $f = \frac{1}{2}h(a, a)\mathrm{d}\tau$.

(iii) Differently to the two 'natural' Galilei connections discussed above in points (i) and (ii), most literature on TNC gravity uses the Galilei connection with torsion

$$T = \hat{v} \otimes \mathrm{d}\tau \tag{6.6a}$$

and local mass torsion representative

$$f = -a(v)\mathrm{d}\tau. \tag{6.6b}$$

There are two ways to see that this choice of f actually satisfies the required transformation behaviour (6.1): either one can check this by a direct calculation; or one checks that with the choice

$$\hat{f} = -\hat{a}(\hat{v})\mathrm{d}\tau = -\hat{\phi}\,\mathrm{d}\tau, \tag{6.6c}$$

the general formula (6.5) for f in terms of $\hat{f}$ gives rise to (6.6b).[2]

We will denote the Galilei connection determined by (6.6) by $\bar{\nabla}$. We may quite easily compute the connection coefficients of $\bar{\nabla}$: its Newton–Coriolis form with respect to $\hat{v}$ is given by $\hat{\Omega} = \hat{f} - \mathrm{d}\hat{a} = -\hat{\phi}\,\mathrm{d}\tau - \mathrm{d}(\hat{\phi}\tau) = -\mathrm{d}\hat{\phi} \wedge \tau - 2\hat{\phi}\,\mathrm{d}\tau$. This implies that $\hat{\Omega}_\nu{}^\rho = \tau_\nu h^{\rho\sigma}\partial_\sigma\hat{\phi} - 2\hat{\phi}(\mathrm{d}\tau)_\nu{}^\rho$. Therefore, according to the classification theorem for Galilei connections (Theorem 2.29), the connection coefficients are

$$\bar{\Gamma}^\rho_{\mu\nu} = \hat{v}^\rho\partial_\mu\tau_\nu + \frac{1}{2}h^{\rho\sigma}(2\partial_{(\mu}\hat{h}_{\nu)\sigma} - \partial_\sigma\hat{h}_{\mu\nu}) + \tau_\mu\tau_\nu h^{\rho\sigma}\partial_\sigma\hat{\phi} - 2\hat{\phi}\tau_{(\mu}(\mathrm{d}\tau)_{\nu)}{}^\rho,$$

$$\tag{6.7a}$$

where $\hat{h}_{\mu\nu}$ denotes the components of the covariant space metric $\underset{\hat{v}}{h}$ with respect to $\hat{v}$. By a direct calculation, one can further show that in terms of the symmetric covariant tensor field $\bar{h} := \underset{\hat{v}}{h} - 2\hat{\phi}\tau \otimes \tau$, this takes the simple form

$$\bar{\Gamma}^\rho_{\mu\nu} = \hat{v}^\rho\partial_\mu\tau_\nu + \frac{1}{2}h^{\rho\sigma}(2\partial_{(\mu}\bar{h}_{\nu)\sigma} - \partial_\sigma\bar{h}_{\mu\nu}). \tag{6.7b}$$

This simple coordinate form is the main reason why $\bar{\nabla}$ is the most widely used Galilei connection in the TNC literature. $\triangle$

---

[2] The choice (6.6c) for $\hat{f}$ allows to generalise this connection to the case of general T (even though this is not commonly discussed in the literature): eq. (6.5) then yields the local mass torsion representative $f = a(T(\cdot, \cdot)) - 2\hat{\phi}\,\mathrm{d}\tau$.

Construction 6.2 In Sect. 3.4, we discussed how to obtain Galilei geometry as the formal Newtonian $c \to \infty$ limit of Lorentzian geometry: we assume that the Lorentzian metric g and its inverse g^{-1} may be expanded as formal power series in c^{-1} as

$$g = -c^2 \tau \otimes \tau + \mathrm{O}(c^0), \tag{6.8a}$$

$$g^{-1} = h + \mathrm{O}(c^{-2}), \tag{6.8b}$$

with τ nowhere-vanishing. According to Lemma 3.48, (M, τ, h) then is a Galilei manifold. Further, we saw in Theorem 3.50 (iv) that $\mathrm{d}\tau = 0$ if and only if the Levi-Civita connection $\overset{g}{\nabla}$ of (M, g) has a regular Newtonian limit.

Now we are going to discuss the c^{-1} expansion of GR *without* assuming $\mathrm{d}\tau = 0$, i.e. allowing the Levi-Civita connection to have *no* regular Newtonian limit. This generalisation was first considered by Van den Bleeken in the article [7].

From Construction 3.49, we know that the Christoffel symbols of g expand as

$$\overset{g}{\Gamma}{}^{\rho}_{\mu\nu} = -c^2 \tau_{(\mu} \, (\mathrm{d}\tau)_{\nu)}{}^{\rho} + \mathrm{O}(c^0). \tag{6.9}$$

Using this, a direct calculation shows (Exercise 6.3) that the Ricci tensor of the Levi-Civita connection $\overset{g}{\nabla}$ of (M, g) satisfies

$$\overset{g}{R}_{\mu\nu} = c^4 \frac{1}{4} \tau_\mu \tau_\nu (\mathrm{d}\tau)_{\rho\sigma} (\mathrm{d}\tau)^{\rho\sigma} + \mathrm{O}(c^2). \tag{6.10}$$

Together with the trace-reversed Einstein equation

$$\overset{g}{R}_{\mu\nu} = \frac{n-1}{n-2} \frac{4\pi G}{c^4} \mathcal{T}_{\mu\nu} \,, \tag{6.11}$$

where $\mathcal{T}_{\mu\nu}$ is the trace-reversed energy–momentum tensor, this shows the following: in a formal c^{-1}-expansion of GR, if the leading-order term of the trace-reversed energy–momentum tensor has order at least c^7, the purely spacelike part $(\mathrm{d}\tau)^{\rho\sigma}$ of $\mathrm{d}\tau$ necessarily vanishes.[3] This is equivalent to $\tau \wedge \mathrm{d}\tau = 0$, the TTNC condition.

In this sense, if the energy–momentum content of a general-relativistic spacetime is not *too* divergent in the Newtonian limit, the Einstein equation guarantees that the limiting Galilei spacetime will allow for an absolute notion of simultaneity. $\triangle$

Exercise 6.3 By inserting the expansion (6.9) of the Christoffel symbols in the coordinate expression for the curvature tensor, show that the Ricci tensor of $\overset{g}{\nabla}$ satisfies (6.10). $\triangle$

---

[3] In fact, (6.10) shows that the order-c^7 term of $\mathcal{T}_{\mu\nu}$ needs to vanish as well.

Remark 6.4 Continuing the c^{-1} expansion of the Einstein equation, higher orders will lead to sourcing of $d\tau \neq 0$ by matter fields [7].

Going beyond the field equation, one may develop a formal c^{-1} expansion of the Einstein–Hilbert action of GR in fully geometric (i.e. coordinate-free) terms [18, 20]. This gives rise to a geometric variational formulation of the post-Newtonian expansion of GR, in the form of a generalisation of standard (i.e. 'torsion-free') Newton–Cartan gravity, including post-Newtonian corrections. For this variational formulation it turns out to be necessary to allow for '(twistless) torsion' $d\tau$ (with $\tau \wedge d\tau = 0$), even if one is in the end only interested in the standard Newton–Cartan case, i.e. in matter that does *not* source $d\tau$. For details, we refer to the original articles [18,20]. $\triangle$

Exercise 6.5 Let (M, g) be a Lorentzian manifold with the metric and inverse metric having formal c^{-1} expansions satisfying Eq. (6.8), and assume that $\tau \wedge d\tau = 0$. Let γ be a worldline in M that is timelike in the Galilei sense. Show that if γ is a geodesic of $\overset{g}{\nabla}$, we have $d\tau = 0$ along γ.

Hint: Using (6.9), expand the geodesic equation in c^{-1}. Use that $\gamma^\mu(\lambda)$ is c-independent by assumption (γ is a curve in M) to conclude that the mixed spacelike-timelike part of $d\tau$ with respect to $\gamma'/\tau(\gamma')$ (which is unit timelike) vanishes along γ. Finally, combine this with $\tau \wedge d\tau = 0$. $\triangle$

Exercise 6.5 shows that even for a 'generalised Newtonian limit' of a general-relativistic spacetime, i.e. one without assuming $d\tau = 0$ from the outset, the consideration of test particle motion automatically leads to standard Newton–Cartan gravity with absolute time (at least along the worldlines of the particles). However, as we will see below in Construction 6.7, this seeming incompatibility of TTNC 'pseudo-Newtonian' limits with particle motion only holds for c-independent worldlines.

Proposition 6.6 *Let (M, g) be a Lorentzian manifold whose metric and inverse metric have formal c^{-1} expansions according to (6.8). Let $\hat{v}$ be the unit timelike vector field on the limiting Galilei manifold (M, τ, h) and $\hat{\phi}$ the function on M that are naturally determined by the next-order terms of the expansions according to Lemma 3.48. Let a be the Bargmann form on (M, τ, h) whose representative with respect to the unit timelike vector field $\hat{v}$ is $\hat{a} = \hat{\phi}\tau.$[4]*

Let further $\bar{\nabla}$ be the Galilei connection on (M, τ, h) with torsion $T = \hat{v} \otimes d\tau$ and mass torsion $f = -a(v)d\tau$, as discussed in Construction 6.1 (iii). Then the following holds:

(i) The Christoffel symbols of g expand as

$$\overset{g}{\Gamma}{}^\rho_{\mu\nu} = -(c^2 h^{\rho\sigma} + m^{\rho\sigma})\tau_{(\mu}(d\tau)_{\nu)\sigma} + \bar{\Gamma}^\rho_{\mu\nu} - \frac{1}{2}\hat{v}^\rho(d\tau)_{\mu\nu} + O(c^{-1}). \quad (6.12)$$

---

[4] Note that according to Theorem 5.25 (iv), we would obtain the same Bargmann form by expanding orthonormal frames in c^{-1}.

(ii) Assuming in addition that $g^{-1} = h + c^{-2}m + \mathrm{O}(c^{-4})$ have no order-c^{-3} term, the order-c^{-1} term of the contraction of the Christoffel symbols with τ vanishes: this contraction is given by

$$\tau_\rho \overset{g}{\Gamma}{}^\rho_{\mu\nu} = \hat{v}^\sigma \tau_{(\mu}(\mathrm{d}\tau)_{\nu)\sigma} + \tau_\rho \bar{\Gamma}^\rho_{\mu\nu} - \frac{1}{2}(\mathrm{d}\tau)_{\mu\nu} + \mathrm{O}(c^{-2})$$

$$= \hat{v}^\sigma \tau_{(\mu}(\mathrm{d}\tau)_{\nu)\sigma} + \partial_{(\mu}\tau_{\nu)} + \mathrm{O}(c^{-2}). \tag{6.13}$$

Proof Comparing the expression for the Christoffel symbols from Construction 3.49 with (6.7a), we directly obtain (6.12).

Further, contracting the calculation in Construction 3.49 with τ_ρ while using $\tau_\rho g^{\rho\sigma} = \tau_\rho(h^{\rho\sigma} + c^{-2}m^{\rho\sigma} + \mathrm{O}(c^{-4})) = -c^{-2}\hat{v}^\sigma + \mathrm{O}(c^{-4})$, we obtain (6.13). $\square$

Construction 6.7 In Exercise 6.5, we saw that in the consideration of Newtonian $c \to \infty$ limits of general-relativistic spacetimes with $\tau \wedge \mathrm{d}\tau = 0$, timelike geodesics are incompatible with $\mathrm{d}\tau \neq 0$. However, as mentioned after this exercise, this incompatibility only arises for c-independent worldlines. Hansen, Hartong, and Obers realised in papers from 2019 and 2020 [19,20] that expanding the geodesic equation for *c-dependent* worldlines allows for solutions in the true TTNC case, where $\tau \wedge \mathrm{d}\tau = 0$, but $\mathrm{d}\tau \neq 0$. In the following, we are going to discuss this expansion of the geodesic equation, focussing on explaining the mathematical setting and discussing the results. A derivation in full detail may be found in Appendix C.[5] Our treatment will be purely in terms of the equations of motion; a discussion of the action principle may be found in the original articles.

As above, we consider a manifold M with a Lorentzian metric g that has a formal power series expansion in c^{-1} as in Lemma 3.48 (i.e. g and g^{-1} satisfy (6.8)). Additionally, we will assume that $g^{-1} = h + c^{-2}m + \mathrm{O}(c^{-4})$ have no order-c^{-3} term, such that we may use (6.13). Finally, we assume the TTNC condition $\tau \wedge \mathrm{d}\tau = 0$. We are interested in the case $\mathrm{d}\tau \neq 0$, such that we take $\mathrm{d}\tau$ to be nowhere-vanishing.

Due to τ satisfying the TTNC condition $\tau \wedge \mathrm{d}\tau = 0$, locally there are smooth functions T and N (an 'integrating factor') such that

$$\tau = N\mathrm{d}T. \tag{6.14}$$

Since τ is nowhere-vanishing, so is N. The spatial leaves integrating the distribution $\ker \tau$ are then the hypersurfaces of constant T. We will write the equations arising from the expansion of the geodesic equation in terms of N. Of course, N is not uniquely determined by (6.14) (and neither is T): N is only unique up to mulitiplication by a function that is spatially constant, i.e. a function constant on any spatial leaf. We will discuss the consequences of this non-uniqueness later.

---

[5] Our notations and conventions are somewhat different from those of the original literature [19,20]: our unit timelike vector fields v^μ and $\hat{v}^\mu$ are their fields $-v^\mu$ and $-\hat{v}^\mu$, respectively; our form ξ_μ (that will appear in $\mathrm{d}\tau$) is their a_μ; and our local representative a_μ of the Bargmann form (which will not explicitly appear in the following discussion) is their m_μ.

Our goal is to expand the geodesic equation

$$\overset{g}{\nabla}_{X'(\lambda)}X'(\lambda) = 0 \tag{6.15}$$

in powers of c^{-1}, for a c-dependent curve $X(\lambda)$ in M. Note that, similar to our discussion of 'c-dependent diffeomorphisms' in Construction 5.28, speaking of a 'c-dependent curve' in M a priori does not make strict mathematical sense: the coordinate representation of a curve in M takes values in $\mathbb{R}^{n+1}$, not in a formal power series ring. However, if c^{-1} were an actual small parameter on which a family of curves $X(\lambda)$ in M depended smoothly, we could expand its components according to

$$X^{\mu}(\lambda) = x^{\mu}(\lambda) + c^{-1}y^{\mu}(\lambda) + O(c^{-2}). \tag{6.16}$$

Geometrically, x is then a (c-independent) curve in M, and y is a vector field along x, i.e. $y(\lambda) \in T_{x(\lambda)}M$ for each value of λ. The expansion (6.16) also makes sense as a formal power series expansion, and this is the starting point for the expansion of the geodesic equation.[6]

For the sake of notational simplicity we will most of the time omit the argument λ. All appearances of X, x, y, their components, etc. are to be evaluated at λ. A prime on any of these objects will denote the derivative with respect to the argument λ.

Before being able to consider the geodesic equation for X, we need to define what we mean by 'evaluation at X' of geometric objects on M, where X is given by the power series expansion (6.16). For any smooth function f on M, this gets a meaning by expanding as if c^{-1} were a small parameter: this yields

$$f\big|_X = f\big|_x + c^{-1}y^{\mu}(\partial_{\mu}f)\big|_x + O(c^{-2}) = f\big|_x + c^{-1}\mathrm{d}f\big|_x(y) + O(c^{-2}). \tag{6.17}$$

Using (6.17), we can now expand the Lorentzian geodesic equation for X (6.15) in powers of c^{-1}. It turns out that in leading orders it is equivalent to the following (for details see Appendix C):

(i) The leading-order curve x is spacelike, i.e. satisfies

$$\tau(x') = 0. \tag{6.18a}$$

(ii) x satisfies the second-order ODE

$$\overset{(n)}{\nabla}_{x'}x' = \frac{1}{2}C^2h(\mathrm{d}N^{-2}, \cdot)\big|_x . \tag{6.18b}$$

---

[6] Equation (6.16) assumes $X(\lambda)$ to have no negative-order terms in c^{-1}. Since the geodesic equation is invariant under affine reparametrisations of the curve, this assumption does not restrict our analysis.

(iii) $\tau(y)$ satisfies the first-order ODE

$$\frac{d}{d\lambda}(\tau(y)) - \tau(y)N^{-1}\Big|_x \frac{d}{d\lambda}N\Big|_x = C \cdot N^{-1}\Big|_x \qquad (6.18c)$$

(which is an equation only for $\tau(y)$ when the solution for x is inserted).

Here $\overset{(n)}{\nabla}$ denotes the Levi-Civita connection of the spatial leaf $(\Sigma, {}^{(n)}h)$ on which x lies, and $C \in \mathbb{R}$ is an integration constant. Up to the considered order(s), the solutions of these equations are uniquely determined by initial values for $x(0), x'(0), \tau(y(0))$. For the physical interpretation, the freedom in the choices of N and C plays a role—this will be discussed below.

The result (6.18b) is an effective equation of motion for a particle of unit mass moving on the Riemannian manifold $(\Sigma, {}^{(n)}h)$ in the potential $V = -\frac{1}{2}C^2 \cdot N^{-2}$ (in the sense of standard Newtonian mechanics). Now consider the expression $g(X', X')$, which is constant along X since it is an (affinely parametrised) geodesic of g. Expanding this in c^{-1}, it may be shown (see Appendix C) to satisfy

$$\frac{1}{2}g(X', X') = \frac{1}{2}{}^{(n)}h(x', x') + V + O(c^{-2}). \qquad (6.19)$$

Therefore, at leading order, $g(X', X')$ being constant is energy conservation for the effective particle motion (6.18b) describing evolution of x. Only effective motions with 'energy' ≤ 0 correspond to physical worldlines of test particles in the Lorentzian spacetime (< 0 for massive particles, $= 0$ for massless ones).

Finally, we will now discuss the impact of the freedom in the choices of N and C on the equations of motion for x and $\tau(y)$. First considering N, we know that it is unique up to multiplication by a spatially constant function. Since x is a spacelike curve, and N is evaluated along x in both the Eqs. (6.18b) for x and (6.18c) for $\tau(y)$, regarding these equations the freedom in the choice of N reduces to freedom of multiplication by a constant. However, when multiplying $N\big|_x$ by some constant, the equations will stay invariant if we scale C by the inverse of this constant. Thus any change in the choice of N may be compensated by a corresponding change in the choice of C. Therefore, fixing N, the discussion of possible different choices for N translates into discussing possible choices for C.

To understand the meaning of C, we consider the velocity of the test particle represented by the Lorentzian geodesic X, as measured by observers moving along the integral curves of $V = -g^{-1}(\tau, \cdot)$ (i.e. by those observers with respect to which we take the Newtonian limit). This velocity is an element of the g-orthogonal complement of V, given by

$$c\,\frac{x'}{C \cdot N^{-1}\big|_x} + O(c^0) \qquad (6.20)$$

(for details see Appendix C). This shows two things. On the one hand, the velocity of the test particle—as seen by the observers with respect to which the Newtonian limit is defined—diverges when taking $c \to \infty$: the 'true TTNC' c^{-1} expansion of the geodesic equation we considered here captures the motion of particles whose velocity is unbounded in the limit. On the other hand, we see that the arbitrary choice of C just sets the velocity scale (for a fixed choice of N): the leading order of the particle's initial velocity in units of c is given by $x'(0)\big/\big(C \cdot N^{-1}\big|_{x(0)}\big)$.

This also captures the invariance of the geodesic equation for X under affine reparametrisations: scaling the parameter of X by some constant corresponds to scaling the parameter of x and y by this constant, as well as scaling $\tau(X') = c^{-1}C \cdot N^{-1}\big|_x + O(c^{-2})$ (see Appendix C) and hence C. On the other hand, one sees directly that such a rescaling of the parameter combined with a rescaling of C maps solutions of the Eqs. (6.18b), (6.18c) to solutions. △

Example 6.8 We consider the c^{-1} expansion of the Schwarzschild metric, but instead of parametrising the Schwarzschild radius as $r_S = \frac{2GM}{c^2}$ in terms of a mass M, we treat it as being c-independent. Physically, this corresponds to taking the formal Newtonian limit for observers that in the limit stay close to the horizon (such that it does not 'shrink to zero size' when taking $c \to \infty$). This limit was first considered in the article [7].

Writing the Schwarzschild metric in terms of r_S, it takes the form

$$g = -\left(1 - \frac{r_S}{r}\right)c^2 dt^2 + \left(1 - \frac{r_S}{r}\right)^{-1} dr^2 + r^2(d\theta^2 + \sin^2\theta\, d\varphi^2), \qquad (6.21\text{a})$$

and its inverse is

$$g^{-1} = -\left(1 - \frac{r_S}{r}\right)^{-1} c^{-2}\partial_t \otimes \partial_t + \left(1 - \frac{r_S}{r}\right)\partial_r \otimes \partial_r$$
$$+ r^{-2}(\partial_\theta \otimes \partial_\theta + \sin^{-2}\theta\, \partial_\varphi \otimes \partial_\varphi). \qquad (6.21\text{b})$$

Comparing this to the Newtonian limit c^{-1} expansions (6.8), we see that these are satisfied with

$$\tau = \sqrt{1 - \frac{r_S}{r}}\, dt, \quad h = \left(1 - \frac{r_S}{r}\right)\partial_r \otimes \partial_r + r^{-2}(\partial_\theta \otimes \partial_\theta + \sin^{-2}\theta\, \partial_\varphi \otimes \partial_\varphi).$$
$$(6.22)$$

Differently to the expansion with $r_S = \frac{2GM}{c^2}$ considered in Example 3.54, treating r_S as c-independent in the Newtonian limit hence yields a Galilei manifold with non-absolute time (which in addition is not spatially flat). Furthermore, we see that the expansion actually *truncates* (at order c^0 for g, and at order c^{-2} for g^{-1}).

For the description of c-dependent Lorentzian geodesics in terms of the limiting TTNC geometry, as discussed in Construction 6.7, we have to write $\tau = N dT$ for functions N and T. In the case at hand, we may choose $T = t$ and

$$N = \sqrt{1 - \frac{r_S}{r}}\,. \qquad (6.23)$$

The solutions of the leading-order equations of motion discussed above, i.e. of $\tau(x') = 0$ and the equation (6.18b) for x, are then *exactly* the spatial projections of causal geodesics in Schwarzschild spacetime [19,20]. △

6.2 String Newton–Cartan Geometry

String Newton–Cartan geometry is a modification of Galilei geometry that enables the description of the movement of strings in spacetime, in such a way that the string worldsheet carries a Lorentzian geometry, while in the directions transversal to the string spacetime is of 'Newtonian' character.

String Newton–Cartan geometry was originally discovered in the article [1] by applying a 'gauging procedure' to the symmetries of a specific large-speed-of-light limit of the Nambu–Goto action for a string in Minkowski spacetime. While this was a purely classical (i.e. non-quantum) consideration, later this was generalised to *quantum* 'non-relativistic string theory', which is a specific large-speed-of-light limit of Poinaré-relativistic string theory with a 'string Galilei symmetry' that had been known since 2001 [17]: it was realised that and how string Newton–Cartan geometry arises as the natural background spacetime geometry to which this 'non-relativistic' quantum string couples [3,16]. Further, it has been studied how to obtain string Newton–Cartan geometry (and string theory on it) by $c \to \infty$ limits of Lorentzian geometry (and string theory on it) [4,23,24]. Null reduction of string theory in Lorentzian backgrounds leads not to string Newton–Cartan geometry, but instead to strings in torsional Newton–Cartan backgrounds [12,22]. These can however be related to string Newton–Cartan geometry [21]. For a thorough discussion of 'non-relativistic' string theory and its relation to string Newton–Cartan as well as torsional Newton–Cartan geometry, we refer to the review [28].

Here, we will only have a look at the very basics of string Newton–Cartan geometry. While most discussions in the literature take a point of view somewhat inspired from quantum field theory and gauge theory, our formulation will be more in what might be called a 'relativist's style', putting more emphasis on the invariant tensor fields characterising the geometry. First, we will discuss the basics of the geometric formalism, in broad parallel to our discussion of Galilei geometry. Note that we will not treat compatible connections; a discussion thereof may be found in the article [6]. Further, we are going to describe the coupling to classical strings in terms of their action principle, and how to obtain string Newton–Cartan geometry as a formal $c \to \infty$ limit of Lorentzian geometry.

We may understand string Newton–Cartan geometry as a generalisation of Galilei geometry without time orientation as discussed in Remark 2.4 (i):

Definition 6.9 A *string Newton–Cartan manifold* is an $(n + 2)$-dimensional differentiable manifold M (with $n \geq 1$) together with

(i) a symmetric covariant degree-2 tensor field $\tilde{\tau} = \tilde{\tau}_{\mu\nu}\mathrm{d}x^\mu \otimes \mathrm{d}x^\nu \in \Gamma(\bigvee^2 T^*M)$ of rank 2, which apart from its degenerate directions has Lorentzian signature, called the *longitudinal metric*, and

(ii) a symmetric contravariant degree-2 tensor field $h = h^{\mu\nu}\partial_\mu \otimes \partial_\nu \in \Gamma(\bigvee^2 TM)$, positive semidefinite of rank n, called the *transversal metric*, such that

(iii) these fields satisfy

$$\tilde{\tau}_{\mu\nu}h^{\nu\rho} = 0. \tag{6.24}$$

A vector $v \in TM$ is *transversal* if $\tilde{\tau}(v, \cdot) = 0$, and *longitudinal* otherwise. A longitudinal vector v is *timelike* if $\tilde{\tau}(v, v) < 0$, *lightlike* or *null* if $\tilde{\tau}(v, v) = 0$, and *spacelike* if $\tilde{\tau}(v, v) > 0$.

As in the Galilei geometry case, h induces a positive-definite bundle metric $^{(n)}h$ on the rank-n distribution $\ker \tilde{\tau}$ of transversal vectors.   ✤

Notation 6.10 When labelling vector fields related to the longitudinal–transversal distinction of tangent directions of a string Newton–Cartan manifold, we will use a *different* index notation than for Galilei frames on Galilei manifolds: we use capital Latin indices as 'longitudinal' indices taking values 0 and 1, and lowercase Latin letters as 'transversal' indices running from 2 to $n + 1$. (We will have no need for indices running through the full range $\{0, \ldots, n + 1\}$.)

We will denote the components of the two-dimensional Minkowski metric in Lorentzian coordinates by η_{AB}, and those of its inverse (which are numerically identical) by η^{AB}; i.e. we have $(\eta_{AB}) = \mathrm{diag}(-1, 1) = (\eta^{AB})$. We will denote the two-dimensional antisymmetric symbol by ε_{AB}, with $\varepsilon_{01} = 1$; i.e. we have

$$(\varepsilon_{AB}) = \begin{pmatrix} 0 & 1 \\ -1 & 0 \end{pmatrix}. \tag{6.25}$$

We will raise the indices of ε_{AB} with η^{AB}, writing for example $\varepsilon^{A}{}_{B} = \eta^{AC}\varepsilon_{CB}$. Note that numerically this means $\varepsilon^{0}{}_{1} = -1 = \varepsilon^{1}{}_{0}$.   ✤

We may now discuss the analogue of unit timelike vector fields in Galilei geometry:

Definition 6.11 Let $(M, \tilde{\tau}, h)$ be a string Newton–Cartan manifold.

(i) Two vector fields $v_A \in \Gamma(U, TM)$ on an open set $U \subset M$ are *(local) orthonormal longitudinal vector fields* if

$$\tilde{\tau}(v_A, v_B) = \eta_{AB} . \tag{6.26}$$

(ii) Analogous to the choice of a unit timelike vector field on a Galilei manifold, on a string Newton–Cartan manifold a choice of orthonormal longitudinal vector fields $v_A \in \Gamma(U, TM)$ determines a unique *transversal projector* $P \in \Gamma(U, TM \otimes T^*M)$ that projects onto $\ker \tilde{\tau}$ and vanishes on $V = \mathrm{span}\{v_A\}$.

Furthermore, it determines a *covariant transversal metric* $\underset{V}{h} \in \Gamma(U, T^*M \otimes T^*M)$, with components $h_{\mu\nu}$, by demanding

$$h_{\mu\nu}v_A^{\nu} = 0, \quad h_{\mu\nu}h^{\nu\rho} = P_{\mu}^{\rho} . \tag{6.27}$$

Finally, a choice of orthonormal longitudinal vector fields v_A determines dual one-forms $\tau^A \in \Omega^1(U)$ by demanding

$$\tau^A(v_B) = \delta_B^A , \quad \tau^A\big|_{\ker \tilde{\tau}} = 0. \tag{6.28}$$

In terms of these, the longitudinal metric can be written as

$$\tilde{\tau} = \eta_{AB}\tau^A \otimes \tau^B,\tag{6.29}$$

and the transversal projector as

$$P = \mathrm{id} - v_A \otimes \tau^A.\tag{6.30}$$

(iii) Fixing a (local) distribution of 'longitudinal spaces' V, the orthonormal longitudinal vector fields $v_A \in \Gamma(U, TM)$ spanning V are unique up to a *(local) longitudinal Lorentz transformation*, i.e. a transformation of the form

$$v_A \mapsto v_B(\Lambda^{-1})^B{}_A\tag{6.31}$$

with a basis change matrix function that takes values in the 2d Lorentz group, $\Lambda \in C^\infty(U, \mathsf{O}(1, 1))$.

(iv) Given two choices v_A and $\tilde{v}_A$ of orthonormal longitudinal vector fields for $(M, \tilde{\tau}, h)$, the $\tilde{v}_A$ arise from v_A by a unique combination of

(a) a *(local) string Galilei boost*

$$v_A \mapsto v_A - w_A\tag{6.32}$$

with transversal vector fields w_A, and

(b) a longitudinal Lorentz transformation.

Local string Galilei boosts are the analogue of local Galilei boosts/Milne boosts for Galilei manifolds in the string Newton–Cartan case.   ✿

Next, we introduce the analogue of a Bargmann form:

Definition 6.12 Let $(M, \tilde{\tau}, h)$ be a string Newton–Cartan manifold. A *string Bargmann form*[7] is an assignment of a two-form $b \in \Omega^2(U)$ to each choice of orthonormal longitudinal vector fields v_A on $U \subset M$ that

(i) is invariant under orientation-preserving longitudinal Lorentz transformations of the v_A,

(ii) flips sign under orientation-reversing longitudinal Lorentz transformations, and

(iii) under local string Galilei boosts $v_A \mapsto v_A - w_A$ transforms as

$$b \mapsto b + \underset{V}{h}(w_0, \cdot) \wedge \tau^1 + \underset{V}{h}(w_1, \cdot) \wedge \tau^0 - \frac{\eta^{AB}}{2}\,^{(n)}h(w_A, w_B)\tau^0 \wedge \tau^1.\tag{6.33a}$$

---

[7] This name is non-standard: in the literature on string Newton–Cartan geometry, this family of fields is usually given no name at all.

Using the (index-raised) antisymmetric symbol, this may be written as

$$b \mapsto b - \varepsilon^A{}_B \underset{V}{h}(w_A, \cdot) \wedge \tau^B - \frac{\eta^{AB}}{2} \, {}^{(n)}h(w_A, w_B)\tau^0 \wedge \tau^1. \tag{6.33b}$$

❀

Remark 6.13 (i) We may show that the different parts of the transformation behaviour of b under changes of v_A are consistent with each other as follows. Starting with orthonormal longitudinal vector fields v_A, first performing a string Galilei boost parametrised by w_A and then a longitudinal Lorentz transformation parametrised by Λ, we end up with $\tilde{v}_A = v_B(\Lambda^{-1})^B{}_A - w_B(\Lambda^{-1})^B{}_A$. On the other hand, we arrive at the same new set of vector fields by first performing a longitudinal Lorentz transformation by Λ to $\hat{v}_A = v_B(\Lambda^{-1})^B{}_A$, and then a string Galilei boost parametrised by $\hat{w}_A = w_B(\Lambda^{-1})^B{}_A$.
Under the first pair of transformations, b transforms according to (6.33b) followed by a multiplication by $\det \Lambda$, i.e. as

$$b \mapsto \det \Lambda \cdot \left(b - \varepsilon^A{}_B \underset{V}{h}(w_A, \cdot) \wedge \tau^B - \frac{\eta^{AB}}{2} \, {}^{(n)}h(w_A, w_B)\tau^0 \wedge \tau^1 \right). \tag{6.34}$$

Under the second pair of transformations, it transforms as

$$b \mapsto \det \Lambda \cdot b - \varepsilon^A{}_B \underset{V}{h}(\hat{w}_A, \cdot) \wedge \hat{\tau}^B - \frac{\eta^{AB}}{2} \, {}^{(n)}h(\hat{w}_A, \hat{w}_B)\hat{\tau}^0 \wedge \hat{\tau}^1, \tag{6.35a}$$

where $\hat{\tau}^A$ are the dual one-forms corresponding to the $\hat{v}_A$. These are given by $\hat{\tau}^A = \Lambda^A{}_B \tau^B$. Writing out several terms, the transformation from above then takes the form

$$b \mapsto \det \Lambda \cdot b - \varepsilon^A{}_B (\Lambda^{-1})^C{}_A \Lambda^B{}_D \underset{V}{h}(w_C, \cdot) \wedge \tau^D$$
$$- \frac{\eta^{AB}(\Lambda^{-1})^C{}_A (\Lambda^{-1})^D{}_B}{2} \, {}^{(n)}h(w_A, w_B)\hat{\tau}^0 \wedge \hat{\tau}^1. \tag{6.35b}$$

To show consistency, we have to show that (6.34) and (6.35b) agree.
First, we note that $\varepsilon_{AB} \Lambda^A{}_C \Lambda^B{}_D = \det \Lambda \cdot \varepsilon_{CD}$. On the one hand, using that Λ is a Lorentz transformation, this implies $\varepsilon^A{}_B(\Lambda^{-1})^C{}_A \Lambda^B{}_D = \varepsilon_{EB}\eta^{AE}(\Lambda^{-1})^C{}_A \Lambda^B{}_D = \varepsilon_{EB}\eta^{CF}\Lambda^E{}_F \Lambda^B{}_D = \eta^{CF} \det \Lambda \cdot \varepsilon_{FD} = \det \Lambda \cdot \varepsilon^C{}_D$, which shows equality of the second terms of (6.34) and (6.35b). On the other hand, due to $\tau^0 \wedge \tau^1 = \frac{1}{2}\varepsilon_{CD}\tau^C \wedge \tau^D$, it implies that $\hat{\tau}^0 \wedge \hat{\tau}^1 = \det \Lambda \cdot \tau^0 \wedge \tau^1$. Together with $\eta^{AB}(\Lambda^{-1})^C{}_A (\Lambda^{-1})^D{}_B = \eta^{CD}$ (since Λ is a Lorentz transformation), this shows equality of the third terms of (6.34) and (6.35b), and we are finished.

(ii) The behaviour of b under longitudinal Lorentz transformations means that b is determined by a local choice of *oriented* longitudinal spaces $V = \mathrm{span}\{v_A\}$. $\triangle$

Notation 6.14 In the following, we are going to discuss worldsheets of strings in a string Newton–Cartan manifold $(M, \tilde{\tau}, h)$, parametrised by smooth 'embedding maps' $\varphi\colon \Sigma \to M$ defined on a 2-dimensional manifold Σ. For index notation for objects on the worldsheet manifold Σ, we will use Greek indices from the beginning of the alphabet (even though no confusion with indices relating to M should be possible). ❀

Definition 6.15 Let $(M, \tilde{\tau}, h)$ be a string Newton–Cartan manifold, and consider a string worldsheet parametrised by $\varphi\colon \Sigma \to M$, where $\dim \Sigma = 2$. On Σ, we consider the induced metric $\varphi^*\tilde{\tau} \in \Gamma(T^*\Sigma \otimes T^*\Sigma)$.

(i) The string's *worldsheet area* is the integral

$$A[\varphi] = \int_\Sigma \mathrm{d}^2\sigma \sqrt{-\det((\varphi^*\tilde{\tau})_{\alpha\beta})} = \int_\Sigma \mathrm{vol}_{\varphi^*\tilde{\tau}} \tag{6.36}$$

of the induced volume element (pseudoform).

(ii) Assume that on $(M, \tilde{\tau}, h)$ we have a string Bargmann form. The string's equation of motion is determined by the *string Newton–Cartan action*[8]

$$S[\varphi] = -\frac{T}{2} \int_\Sigma \left(\mathrm{vol}_{\varphi^*\tilde{\tau}} \left((\varphi^*\tilde{\tau})^{-1} \right)^{\alpha\beta} (\varphi^* \underset{V}{h})_{\alpha\beta} + 2\varphi^* o_V\, \varphi^* b \right). \tag{6.37}$$

This is only defined for embeddings φ such that the induced metric is non-degenerate, i.e. for strings moving such that their worldsheet is longitudinal (i.e. the tangent vectors of $\varphi(\Sigma)$ are longitudinal). Here T is the string tension, and we have made a (local) choice of orthonormal longitudinal vector fields v_A, which determine a local choice of oriented longitudinal spaces $V = \mathrm{span}\{v_A\}$, the covariant transversal metric $\underset{V}{h}$, and the representative b of the string Bargmann form. Further, $o_V = \mathrm{sgn}(\tau^0 \wedge \tau^1)$ denotes the local orientation on longitudinal spaces induced by that of V. Since the tangent vectors of Σ map to longitudinal ones, $\varphi^* o_V$ locally defines an orientation on Σ (pointwise understood as a map {ordered bases of $T_\sigma\Sigma$} $\to \{\pm 1\}$). Hence the product $\varphi^* o_V\, \varphi^* b$ is a 2-pseudoform on Σ, which can be integrated even if Σ is non-orientable.[9]

---

[8] This is what might be called the 'Nambu–Goto form' of the action. In the literature, a 'Polyakov form' is often considered, which introduces several additional fields, including an independent worldsheet metric. Integrating out these additional fields, the Polyakov reduces to the Nambu–Goto form.

[9] A *k-pseudoform* on a differentiable manifold N is an assignment to each oriented open subset $U \subset N$ of a k-form on U such that these k-forms flip sign under change of local orientation, and are compatible with restriction to smaller open sets. Locally, a k-pseudoform can be written as a product of a (local) orientation and a k-form; and conversely, any globally defined object which can locally be expressed in this way is a k-pseudoform. A k-pseudoform may be characterised

One may show that the integrand of the action is in fact independent of the choice of orthonormal longitudinal vector fields v_A: it is clearly invariant under orientation-preserving longitudinal Lorentz transformations of the v_A. Further, due to both o_V and b flipping their signs under orientation-reversing longitudinal Lorentz transformations of the v_A, the pseudoform $\varphi^* o_V\, \varphi^* b$ is invariant under these, such that the action's integrand is invariant as well. Showing invariance under string Galilei boosts requires a somewhat lengthy calculation, which we omit for the sake of brevity.   ❀

Finally, we are going to discuss how to obtain a string Newton–Cartan manifold as a specific kind of formal $c \to \infty$ limit of a Lorentzian manifold. Differently to the Newtonian limit leading to Galilei geometry, in the limit leading to string Newton–Cartan geometry the Lorentzian lightcones 'open up' as $c \to \infty$ only in the transversal directions, such that the longitudinal directions keep their Lorentzian character. We will formulate the limiting construction in terms of orthonormal frames, in direct parallel to that for Galilei geometry with a Bargmann form discussed in Theorem 5.25.

Construction 6.16 Let (M, g) be an $(n + 2)$-dimensional Lorentzian manifold, and let (E_A, E_a) be a local orthonormal frame with dual frame (E^A, E^a), such that the metric and inverse metric can be written as

$$g = \eta_{AB} E^A \otimes E^B + \delta_{ab} E^a \otimes E^b, \quad g^{-1} = \eta^{AB} E_A \otimes E_B + \delta^{ab} E_a \otimes E_b. \quad (6.38)$$

Assume that the frame and dual frame may be expanded as formal power series in c^{-1} as

$$E^A = c\tau^A + c^{-1} a^A + O(c^{-3}), \qquad E^a = e^a + O(c^{-2}), \qquad (6.39a)$$

$$E_A = c^{-1} v_A + O(c^{-3}), \qquad E_a = e_a + O(c^{-2}) \qquad (6.39b)$$

for nowhere-vanishing local one-forms τ^A.[10]
    Then

$$\tilde{\tau} = \eta_{AB} \tau^A \otimes \tau^B, \quad h = \delta^{ab} e_a \otimes e_b \qquad (6.40)$$

---

by its components with respect to ordered bases, which under basis change transform as those of a k-form, apart from an additional minus sign under orientation-reversing basis changes. On any n-dimensional manifold, there is a well-defined notion of integration of n-pseudoforms (with compact support), without any need for orientability. Details on pseudoforms and their integration theory may be found in Sects. 2.8 and 3.4 of the textbook [11].

[10] For the sake of simplicity, following the literature [23,24], we assume the leading order and next to leading order of the frame and dual frame fields to differ by a factor of c^{-2}. Note, however, that this is a stronger assumption than the one we made in Theorem 5.25 regarding the limit from Lorentzian to Galilei geometry with a Bargmann form; and probably it may be relaxed.

define a string Newton–Cartan manifold $(M, \tilde{\tau}, h)$, and the v_A are local orthonormal longitudinal vector fields. The covariant transversal metric with respect to the v_A can be written as

$$h_{V} = \delta_{ab}\mathrm{e}^a \otimes \mathrm{e}^b. \tag{6.41}$$

String Galilei boosts of the v_A arise from local Lorentz transformations

$$(\mathrm{E}_A, \mathrm{E}_a) \mapsto (\mathrm{E}_A, \mathrm{E}_a) \cdot \Phi^{-1}, \quad \Phi \in C^\infty(U, \mathrm{O}(1, n+1)), \tag{6.42}$$

acting in transversal–longitudinal planes: for Φ a Lorentz boost with purely transversal velocity k^a, the transformation (6.42) induces a string Galilei boost with parameter vector fields $w_0 = k^a \mathrm{e}_a$, $w_1 = 0$. Similarly, if Φ is a rotation from the spacelike longitudinal direction 1 towards a unit transversal direction n^a with angle $|k|/c$, i.e.

$$\Phi^0{}_0 = 1, \quad \Phi^0{}_1 = \Phi^0{}_a = \Phi^1{}_0 = \Phi^a{}_0 = 0, \tag{6.43a}$$

$$\Phi^1{}_1 = \cos(|k|/c), \quad \Phi^a{}_1 = \sin(|k|/c)n^a = -\delta^{ab}\Phi^1{}_b, \tag{6.43b}$$

$$\Phi^a{}_b = \delta^a_b + (\cos(|k|/c) - 1)n^a n_b, \tag{6.43c}$$

the local Lorentz transformation (6.42) induces a string Galilei boost parametrised by $w_0 = 0$, $w_1 = |k|n^a \mathrm{e}_a$.

The induced transformations on the subleading one-forms a^A are then such that the two-form

$$b = \varepsilon_{AB}\tau^A \wedge a^B = \tau^0 \wedge a^1 - \tau^1 \wedge a^0 \tag{6.44}$$

transforms correctly under string Galilei boosts such as to define a string Bargmann form.   △

Exercise 6.17 (*String Newton–Cartan expansion of the Nambu–Goto action*) We consider the Nambu–Goto action for a string in a Lorentzian manifold (M, g), parametrised by $\varphi \colon \Sigma \to M$. It is $-T$ (the negative of the string tension) times the string's worldsheet area in the metric induced by g, i.e.

$$S_{\mathrm{NG}}[\varphi] = -T \int_\Sigma \mathrm{vol}_{\varphi^* g} = -T \int_\Sigma \mathrm{d}^2\sigma \sqrt{-\det((\varphi^* g)_{\alpha\beta})}. \tag{6.45}$$

Show that in a c^{-1} expansion of the Lorentzian metric leading to string Newton–Cartan geometry as in Construction 6.16, the Nambu–Goto action expands as

$$S_{\mathrm{NG}}[\varphi] = -Tc^2 A[\varphi]$$

$$- \frac{T}{2} \int_\Sigma \mathrm{vol}_{\varphi^* \tilde{\tau}} \left((\varphi^* \tilde{\tau})^{-1}\right)^{\alpha\beta}\left((\varphi^* h_{V})_{\alpha\beta} + 2\eta_{AB}(\varphi^* \tau^A)_\alpha (\varphi^* a^B)_\beta\right)$$

$$+ \mathrm{O}(c^{-2}). \tag{6.46}$$

Hint: Insert $g = \eta_{AB}E^A \otimes E^B + \delta_{ab}E^a \otimes E^b$, *and use* (6.39). *For expanding the determinant, use Jacobi's formula for the derivative of the determinant,* $\mathrm{d}\det A = \det A \cdot \mathrm{tr}(A^{-1}\mathrm{d}A)$. $\qquad\qquad\qquad\qquad\qquad\qquad\qquad\qquad\qquad\qquad\qquad\qquad\triangle$

Remark 6.18 Similar to the discussion of c-dependent particle worldlines in TTNC gravity (Construction 6.7), one may generalise this discussion of the c^{-1} expansion of the dynamics of a string in Lorentzian geometry by allowing the 'embedding map' φ to be c-dependent as well. This is discussed in the articles [23,24]. $\qquad\triangle$

Construction 6.19 The second term in the c^{-1} expansion (6.46) of the Nambu–Goto action is actually equal to the string Newton–Cartan action (6.37), i.e.

$$S_{\mathrm{NG}}[\varphi] = -Tc^2 A[\varphi] + S[\varphi] + \mathrm{O}(c^{-2}). \tag{6.47}$$

In order to show this, it is sufficient to show that

$$\mathrm{vol}_{\varphi^*\tilde{\tau}}\left((\varphi^*\tilde{\tau})^{-1}\right)^{\alpha\beta}\eta_{AB}(\varphi^*\tau^A)_\alpha(\varphi^*a^B)_\beta = \varphi^*o_V\,\varphi^*b. \tag{6.48}$$

This works as follows.

On an n-dimensional manifold N with pseudo-Riemannian metric m, the Hodge star operator $\star$ with respect to m maps k-forms to $(n-k)$-pseudoforms. It is defined by requiring that for any two k-forms $\alpha, \beta \in \Omega^k(N)$ we have

$$\alpha \wedge \star\beta = m^{-1}(\alpha, \beta)\,\mathrm{vol}_m\ , \tag{6.49}$$

where m^{-1} has been extended to k-forms in the natural way, i.e.

$$m^{-1}(\alpha, \beta) = \frac{1}{k!}m^{\mu_1\nu_1}\ldots m^{\nu_1\mu_k}\alpha_{\mu_1\ldots\mu_k}\beta_{\nu_1\ldots\nu_k}\ , \tag{6.50}$$

and vol_m is the volume element (n-pseudoform) induced by m.

We now apply this to the worldsheet manifold Σ with its induced Lorentzian metric $\varphi^*\tilde{\tau}$. This can be written as $\varphi^*\tilde{\tau} = \eta_{AB}\varphi^*\tau^A \otimes \varphi^*\tau^B$, implying that the one-forms $\varphi^*\tau^A$ are orthonormal. Therefore, the induced volume element can be written as

$$\mathrm{vol}_{\varphi^*\tilde{\tau}} = \varphi^*o_V\,\varphi^*\tau^0 \wedge \varphi^*\tau^1. \tag{6.51}$$

Hence, the Hodge star operator with respect to $\varphi^*\tilde{\tau}$ acts on these forms as

$$\star\varphi^*\tau^0 = \varphi^*o_V\,\varphi^*\tau^1, \quad \star\varphi^*\tau^1 = \varphi^*o_V\,\varphi^*\tau^0. \tag{6.52a}$$

Using the (index-raised) antisymmetric symbol, this may be written as

$$\star\varphi^*\tau^A = -\varepsilon^A{}_B\varphi^*o_V\,\varphi^*\tau^B. \tag{6.52b}$$

In terms of the Hodge operator, we can rewrite the left-hand side of (6.48) as

$$\mathrm{vol}_{\varphi^*\tilde\tau}\left((\varphi^*\tilde\tau)^{-1}\right)^{\alpha\beta}\eta_{AB}(\varphi^*\tau^A)_\alpha(\varphi^*a^B)_\beta = \mathrm{vol}_{\varphi^*\tilde\tau}\,\eta_{AB}(\varphi^*\tilde\tau)^{-1}(\varphi^*a^B,\varphi^*\tau^A)$$
$$= \eta_{AB}\varphi^*a^B \wedge \star\varphi^*\tau^A. \tag{6.53}$$

Inserting (6.52b), this becomes

$$\eta_{AB}\varphi^*a^B \wedge \star\varphi^*\tau^A = \eta_{AB}\varphi^*a^B \wedge (-\varepsilon^A{}_C)\varphi^*o_V\,\varphi^*\tau^C$$
$$= -\varphi^*o_V\,\varepsilon_{BC}\varphi^*a^B \wedge \varphi^*\tau^C$$
$$= -\varphi^*o_V\,\varphi^*(\varepsilon_{BC}a^B \wedge \tau^C). \tag{6.54}$$

Using the identification (6.44) of b in the c^{-1} expansion leading from Lorentzian to string Newton–Cartan geometry, this expression is equal to $\varphi^*o_V\,\varphi^*b$, finishing the proof of (6.48). △

References

1. Andringa, R., Bergshoeff, E., Gomis, J., de Roo, M.: 'Stringy' Newton–Cartan gravity. Class. Quantum Gravity **29**(23), 235020 (2012). https://doi.org/10.1088/0264-9381/29/23/235020
2. Bekaert, X., Morand, K.: Connections and dynamical trajectories in generalised Newton-Cartan gravity I. An intrinsic view. J. Math. Phys. **57**(2), 022507 (2016). https://doi.org/10.1063/1.4937445
3. Bergshoeff, E., Gomis, J., Yan, Z.: Nonrelativistic string theory and T-duality. J. High Energy Phys. **2018**(11), 133 (2018). https://doi.org/10.1007/JHEP11(2018)133
4. Bergshoeff, E.A., Gomis, J., Rosseel, J., Şimşek, C., Yan, Z.: String theory and string Newton–Cartan geometry. J. Phys. A: Math. Theor. **53**(1), 014001 (2019). https://doi.org/10.1088/1751-8121/ab56e9
5. Bergshoeff, E.A., Hartong, J., Rosseel, J.: Torsional Newton–Cartan geometry and the Schrödinger algebra. Class. Quantum Gravity **32**(13), 135017 (2015). https://doi.org/10.1088/0264-9381/32/13/135017
6. Bergshoeff, E.A., van Helden, K., Lahnsteiner, J., Romano, L., Rosseel, J.: Generalized Newton–Cartan geometries for particles and strings. Class. Quantum Gravity **40**(7), 075010 (2023). https://doi.org/10.1088/1361-6382/acbe8c
7. Van den Bleeken, D.: Torsional Newton–Cartan gravity from the large c expansion of general relativity. Class. Quantum Gravity **34**(18), 185004 (2017). https://doi.org/10.1088/1361-6382/aa83d4
8. Christensen, M.H., Hartong, J., Obers, N.A., Rollier, B.: Boundary stress-energy tensor and Newton-Cartan geometry in Lifshitz holography. J. High Energy Phys. **2014**(1), 57 (2014). https://doi.org/10.1007/JHEP01(2014)057
9. Christensen, M.H., Hartong, J., Obers, N.A., Rollier, B.: Torsional Newton-Cartan geometry and Lifshitz holography. Phys. Rev. D **89**(6), 061901 (2014). https://doi.org/10.1103/PhysRevD.89.061901
10. Ergen, M., Hamamcı, E., Van den Bleeken, D.: Oddity in nonrelativistic, strong gravity. Eur. Phys. J. C **80**(6), 563 (2020). https://doi.org/10.1140/epjc/s10052-020-8112-6
11. Frankel, T.: The Geometry of Physics, 3rd edn. Cambridge University Press, Cambridge (2011). https://doi.org/10.1017/CBO9781139061377

12. Gallegos, A.D., Gürsoy, U., Zinnato, N.: Torsional Newton Cartan gravity from non-relativistic strings. J. High Energy Phys. **2020**(9), 172 (2020). https://doi.org/10.1007/JHEP09(2020)172
13. Geracie, M., Prabhu, K., Roberts, M.M.: Curved non-relativistic spacetimes, Newtonian gravitation and massive matter. J. Math. Phys. **56**(10), 103505 (2015). https://doi.org/10.1063/1.4932967
14. Geracie, M., Prabhu, K., Roberts, M.M.: Fields and fluids on curved non-relativistic spacetimes. J. High Energy Phys. **2015**(8), 42 (2015). https://doi.org/10.1007/JHEP08(2015)042
15. Geracie, M., Son, D.T., Wu, C., Wu, S.F.: Spacetime symmetries of the quantum Hall effect. Phys. Rev. D **91**(4), 045030 (2015). https://doi.org/10.1103/PhysRevD.91.045030
16. Gomis, J., Oh, J., Yan, Z.: Nonrelativistic string theory in background fields. J. High Energy Phys. **2019**(10), 101 (2019). https://doi.org/10.1007/JHEP10(2019)101
17. Gomis, J., Ooguri, H.: Nonrelativistic closed string theory. J. Math. Phys. **42**(7), 3127–3151 (2001). https://doi.org/10.1063/1.1372697
18. Hansen, D., Hartong, J., Obers, N.A.: Action principle for Newtonian gravity. Phys. Rev. Lett. **122**(6), 061106 (2019). https://doi.org/10.1103/PhysRevLett.122.061106
19. Hansen, D., Hartong, J., Obers, N.A.: Gravity between Newton and Einstein. Int. J. Mod. Phys. D **28**(14), 1944010 (2019). https://doi.org/10.1142/S0218271819440103
20. Hansen, D., Hartong, J., Obers, N.A.: Non-relativistic gravity and its coupling to matter. J. High Energy Phys. **2020**(6), 145 (2020). https://doi.org/10.1007/JHEP06(2020)145
21. Harmark, T., Hartong, J., Menculini, L., Obers, N.A., Oling, G.: Relating non-relativistic string theories. J. High Energy Phys. **2019**(11), 71 (2019). https://doi.org/10.1007/JHEP11(2019)071
22. Harmark, T., Hartong, J., Obers, N.A.: Nonrelativistic strings and limits of the AdS/CFT correspondence. Phys. Rev. D **96**(8), 086019 (2017). https://doi.org/10.1103/PhysRevD.96.086019
23. Hartong, J., Have, E.: Nonrelativistic expansion of closed bosonic strings. Phys. Rev. Lett. **128**(2), 021602 (2022). https://doi.org/10.1103/PhysRevLett.128.021602
24. Hartong, J., Have, E.: Nonrelativistic approximations of closed bosonic string theory. J. High Energy Phys. **2023**(2), 153 (2023). https://doi.org/10.1007/JHEP02(2023)153
25. Hartong, J., Kiritsis, E., Obers, N.A.: Lifshitz space-times for Schrödinger holography. Phys. Lett. B **746**, 318–324 (2015). https://doi.org/10.1016/j.physletb.2015.05.010
26. Hartong, J., Obers, N.A.: Hořava-Lifshitz gravity from dynamical Newton-Cartan geometry. J. High Energy Phys. **2015**(7), 155 (2015). https://doi.org/10.1007/JHEP07(2015)155
27. Hartong, J., Obers, N.A., Oling, G.: Review on non-relativistic gravity. Front. Phys. **11**, 1116888 (2023). https://doi.org/10.3389/fphy.2023.1116888
28. Oling, G., Yan, Z.: Aspects of nonrelativistic strings. Front. Phys. **10**, 832271 (2022). https://doi.org/10.3389/fphy.2022.832271
29. Son, D.T.: Newton-Cartan geometry and the quantum Hall effect (2013). https://doi.org/10.48550/arXiv.1306.0638. arXiv:1306.0638

Semidirect Products of Lie Groups

A

Abstract

Here, we briefly introduce the basics of semidirect products of Lie groups, semidirect sums of Lie algebras, and their relation.

Definition A.1 Let H, N be Lie groups and $\rho \colon H \to \mathrm{Aut}(N), h \mapsto \rho_h$ a group homomorphism such that the joint map $H \times N \ni (h, n) \mapsto \rho_h(n) \in N$ is smooth[1] (here $\mathrm{Aut}(N)$ is the group of Lie group automorphisms of N, i.e. isomorphisms $N \to N$). The *semidirect product* of H and N with respect to ρ, denoted $H \ltimes_\rho N$, is the Lie group whose underlying manifold is the product $H \times N$ and whose group operation is

$$(h, n)(\tilde{h}, \tilde{n}) = (h\tilde{h}, n\rho_h(\tilde{n})), \tag{A.1}$$

i.e. in the group operation 'H acts on N via ρ'. ✾

That this operation really is associative is guaranteed by ρ being a homomorphism and taking values in automorphisms. The neutral element of $H \ltimes_\rho N$ is (e_H, e_N) (where e_H and e_N are the neutral elements of H and N, respectively), and inverses are given by

$$(h, n)^{-1} = (h^{-1}, \rho_{h^{-1}}(n^{-1})) \tag{A.2}$$

(Exercise A.2). If the homomorphism by which H acts on N is clear from context, we will omit it from the notation.

Exercise A.2 Let H, N be groups and $\rho \colon H \to \mathrm{Aut}(N)$ a group homomorphism. Show that the group operation on the semidirect product $H \ltimes_\rho N$ is associative, (e_H, e_N) is its neutral element, and inverses are given by (A.2). △

---

[1] If $\mathrm{Aut}(N)$ is a Lie group itself, this is equivalent to ρ being smooth. However, if the group of connected components of N is not finitely generated, $\mathrm{Aut}(N)$ need not be a Lie group.

© The Editor(s) (if applicable) and The Author(s), under exclusive license to Springer Nature Switzerland AG 2026

P. K. Schwartz, *Newton–Cartan Gravity*, Lecture Notes in Physics 1044, https://doi.org/10.1007/978-3-032-03967-5

Both H and N embed as subgroups into $H \ltimes N$ via the maps

$$H \ni h \mapsto (h, e_N) \in H \ltimes N, \tag{A.3a}$$

$$N \ni n \mapsto (e_H, n) \in H \ltimes N. \tag{A.3b}$$

N is a normal subgroup of $H \ltimes N$, with the quotient being canonically isomorphic to H. This means we have a short exact sequence

$$1 \to N \to H \ltimes N \to H \to 1. \tag{A.4}$$

In general, H as a subgroup is not normal.

An analogous construction exists for Lie algebras:

Definition A.3 Let $\mathfrak{h}, \mathfrak{n}$ be Lie algebras and $\hat{\rho} \colon \mathfrak{h} \to \mathsf{Der}(\mathfrak{n})$ a Lie algebra homomorphism (here $\mathsf{Der}(\mathfrak{n})$ is the Lie algebra of derivations on $\mathfrak{n}$, i.e. linear maps $f \colon \mathfrak{n} \to \mathfrak{n}$ satisfying a 'Leibniz rule' with respect to the Lie bracket, $f([Y, \tilde{Y}]) = [f(Y), \tilde{Y}] + [Y, f(\tilde{Y})]$). The *semidirect sum* of $\mathfrak{h}$ and $\mathfrak{n}$ with respect to $\hat{\rho}$, denoted $\mathfrak{h} \oplus_{\hat{\rho}} \mathfrak{n}$, is the Lie algebra whose underlying vector space is the direct sum $\mathfrak{h} \oplus \mathfrak{n}$ and whose Lie bracket is

$$[(X, Y), (\tilde{X}, \tilde{Y})] = ([X, \tilde{X}], [Y, \tilde{Y}] + \hat{\rho}_X(\tilde{Y}) - \hat{\rho}_{\tilde{X}}(Y)), \tag{A.5}$$

i.e. in the Lie bracket '$\mathfrak{h}$ acts on $\mathfrak{n}$ via $\hat{\rho}$'.[2]   ✤

That this bracket really satisfies the Jacobi identity is guaranteed by $\hat{\rho}$ being a homomorphism and taking values in derivations (Exercise A.4). As for semidirect products of groups, if the homomorphism is clear from context, we will omit it from the notation.

Exercise A.4 Let $\mathfrak{h}, \mathfrak{n}$ be Lie algebras and $\hat{\rho} \colon \mathfrak{h} \to \mathsf{Der}(\mathfrak{n})$ a Lie algebra homomorphism. Show that the Lie bracket on the semidirect sum $\mathfrak{h} \oplus_{\hat{\rho}} \mathfrak{n}$ satisfies the Jacobi identity.   △

Both $\mathfrak{h}$ and $\mathfrak{n}$ embed as Lie subalgebras into $\mathfrak{h} \oplus \mathfrak{n}$ via the maps

$$\mathfrak{h} \ni X \mapsto (X, 0) \in \mathfrak{h} \oplus \mathfrak{n}, \tag{A.6a}$$

$$\mathfrak{h} \ni Y \mapsto (0, Y) \in \mathfrak{h} \oplus \mathfrak{n}. \tag{A.6b}$$

$\mathfrak{n}$ is an ideal of $\mathfrak{h} \oplus \mathfrak{n}$, with the quotient being canonically isomorphic to $\mathfrak{h}$. This means we have a short exact sequence

---

[2] Note that the notation with $\oplus$ is non-standard. Most literature simply uses the direct sum symbol, but we want to take account of the non-trivial Lie bracket structure in the notation, and have it reflect the 'direction' of the action ('from left to right'), similar to the standard $\ltimes$ notation for semidirect products.

$$0 \to \mathfrak{n} \to \mathfrak{h} \oplus \mathfrak{n} \to \mathfrak{h} \to 0. \tag{A.7}$$

In general, $\mathfrak{h}$ as a subalgebra is not an ideal. Semidirect products of Lie groups and semidirect sums of Lie algebras are closely related. To specify this, we need the following.

Construction A.5 Let H, N be Lie groups and $\rho \colon H \to \mathsf{Aut}(N)$ a group homomorphism such that the joint map $H \times N \ni (h, n) \mapsto \rho_h(n) \in N$ is smooth. For each $h \in H$, we consider the differential of $\rho_h \colon N \to N$ at the neutral element $e_N \in N$, which we denote by

$$\dot{\rho}_h := \mathrm{D}(\rho_h)|_{e_N} \colon T_{e_N} N \to T_{e_N} N. \tag{A.8}$$

Since ρ_N is a Lie group automorphism, this differential is a Lie algebra automorphism $\dot{\rho}_h \in \mathsf{Aut}(\mathfrak{n})$. By the chain rule, the fact that ρ is a Lie group homomorphism translates into

$$\dot{\rho} \colon H \to \mathsf{Aut}(\mathfrak{n}), h \mapsto \dot{\rho}_h = \mathrm{D}(\rho_h)|_{e_N} \tag{A.9}$$

being a Lie group homomorphism.

Now of *this* homomorphism, we again take the differential at the neutral element $e_H \in H$, which is a Lie algebra homomorphism

$$\dot{\rho}' := \mathrm{D}\dot{\rho}|_{e_H} \colon \mathfrak{h} \to \mathsf{Lie}(\mathsf{Aut}(\mathfrak{n})). \tag{A.10}$$

But the Lie algebra of $\mathsf{Aut}(\mathfrak{n})$ consists of derivations on $\mathfrak{n}$ (if you want, check this as an exercise), so we have seen how a Lie group homomorphism $\rho \colon H \to \mathsf{Aut}(N)$ induces a Lie algebra homomorphism

$$\dot{\rho}' \colon \mathfrak{h} \to \mathsf{Der}(\mathfrak{n}). \tag{A.11}$$

$$\triangle$$

Proposition A.6 *Let H, N be Lie groups and $\rho \colon H \to \mathsf{Aut}(N)$ a group homomorphism such that the joint map $H \times N \ni (h, n) \mapsto \rho_h(n) \in N$ is smooth.*

(i) The Lie algebra of the semidirect product $H \ltimes_\rho N$ is the semidirect sum $\mathfrak{h} \oplus_{\dot{\rho}'} \mathfrak{n}$, where $\dot{\rho}' \colon \mathfrak{h} \to \mathsf{Der}(\mathfrak{n})$ is the Lie algebra homomorphism induced by ρ (according to Construction A.5).

(ii) The adjoint representation of $H \ltimes_\rho N$ is given by

$$\mathrm{Ad}_{(h,n)}(X, Y) = (\mathrm{Ad}_h(X), \mathrm{Ad}_n(\dot{\rho}_h(Y)) + \sigma_n(\mathrm{Ad}_h(X))) \tag{A.12}$$

for $(h, n) \in H \ltimes_\rho N$ and $(X, Y) \in \mathfrak{h} \oplus_{\dot{\rho}'} \mathfrak{n}$, where $\dot{\rho} \colon H \to \mathsf{Aut}(\mathfrak{n})$ is the homomorphism induced by ρ, and $\sigma_n \colon \mathfrak{h} \to \mathfrak{n}$ is the differential of the map $H \to N, h \mapsto n\rho_h(n^{-1})$ at the neutral element e_H.

Proof As a vector space, we know that the Lie algebra of the semidirect product is $\mathrm{Lie}(H \ltimes_\rho N) = T_{(e_H, e_N)}(H \ltimes_\rho N) = T_{e_H} H \oplus T_{e_N} N = \mathfrak{h} \oplus \mathfrak{n}$. First, we will compute the adjoint representation. Let $h \in H, n \in N$. We denote by $\alpha_h \colon H \to H$ the conjugation by h, i.e. $\alpha_h(\tilde{h}) = h\tilde{h}h^{-1}$; similarly α_n is conjugation by n. The conjugation map by $(h, n) \in H \ltimes_\rho N$ is then given by

$$
\begin{aligned}
\alpha_{(h,n)}(\tilde{h}, \tilde{n}) &= (h, n) \cdot (\tilde{h}, \tilde{n}) \cdot (h, n)^{-1} \\
&= (h\tilde{h}, n\rho_h(\tilde{n})) \cdot (h^{-1}, \rho_{h^{-1}}(n^{-1})) \\
&= \left(h\tilde{h}h^{-1}, n\rho_h(\tilde{n})\rho_{h\tilde{h}}(\rho_{h^{-1}}(n^{-1})) \right) \\
&= \left(\alpha_h(\tilde{h}), n\rho_h(\tilde{n})\rho_{\alpha_h(\tilde{h})}(n^{-1}) \right).
\end{aligned}
\tag{A.13}
$$

Now, in order to compute the adjoint representation

$$
\mathrm{Ad}_{(h,n)} = \mathrm{D}(\alpha_{(h,n)})\big|_{(e_H, e_N)} \colon \mathfrak{h} \oplus \mathfrak{n} \to \mathfrak{h} \oplus \mathfrak{n},
\tag{A.14}
$$

consider $(X, Y) \in \mathfrak{h} \oplus \mathfrak{n}$ and let $\tilde{h}(t), \tilde{n}(t)$ be curves in H and N respectively such that $\dot{\tilde{h}}(0) = X, \dot{\tilde{n}}(0) = Y$. We then may compute

$$
\begin{aligned}
\mathrm{Ad}_{(h,n)}(X, Y) &= \frac{\mathrm{d}}{\mathrm{d}t}\alpha_{(h,n)}(\tilde{h}(t), \tilde{n}(t))\bigg|_{t=0} \\
&= \frac{\mathrm{d}}{\mathrm{d}t}\alpha_{(h,n)}(\tilde{h}(t), e_N)\bigg|_{t=0} + \frac{\mathrm{d}}{\mathrm{d}t}\alpha_{(h,n)}(e_H, \tilde{n}(t))\bigg|_{t=0} \\
&= \frac{\mathrm{d}}{\mathrm{d}t}\left(\alpha_h(\tilde{h}(t)), n\rho_{\alpha_h(\tilde{h}(t))}(n^{-1}) \right)\bigg|_{t=0} \\
&\quad + \frac{\mathrm{d}}{\mathrm{d}t}\left(e_H, n\rho_h(\tilde{n}(t))n^{-1} \right)\bigg|_{t=0} \\
&= \frac{\mathrm{d}}{\mathrm{d}t}\left(\alpha_h(\tilde{h}(t)), \beta_n(\alpha_h(\tilde{h}(t))) \right)\bigg|_{t=0} + \frac{\mathrm{d}}{\mathrm{d}t}(e_H, \alpha_n(\rho_h(\tilde{n}(t))))\bigg|_{t=0},
\end{aligned}
\tag{A.15}
$$

where we introduced the notation $\beta_n(\hat{h}) := n\rho_{\hat{h}}(n^{-1})$. Denoting the differential of this map by $\sigma_n := \mathrm{D}\beta_n|_{e_H} \colon \mathfrak{h} \to \mathfrak{n}$ as in the statement of the proposition, we obtain

$$
\begin{aligned}
\mathrm{Ad}_{(h,n)}(X, Y) &= \left(\mathrm{Ad}_h(\dot{\tilde{h}}(0)), \sigma_n(\mathrm{Ad}_h(\dot{\tilde{h}}(0))) \right) + \left(0, \mathrm{Ad}_n(\dot{\rho}_h(\dot{\tilde{n}}(0))) \right) \\
&= (\mathrm{Ad}_h(X), \mathrm{Ad}_n(\dot{\rho}_h(Y)) + \sigma_n(\mathrm{Ad}_h(X))).
\end{aligned}
\tag{A.16}
$$

To prove that the Lie algebra is the semidirect sum, we may now use this result. This is covered in Exercise A.7. ∎

Exercise A.7 In this exercise, we are going to finish the proof of Proposition A.6, i.e. show that the Lie algebra of $H \ltimes_\rho N$ is indeed the semidirect sum $\mathfrak{h} \oplus_{\dot\rho'} \mathfrak{n}$. We already know that *as a vector space*, the Lie algebra is $\mathfrak{h} \oplus \mathfrak{n}$; therefore we have to compute the Lie bracket.

For this, we may use the already proved result (A.12) on the adjoint representation Ad of the group:[3] the Lie bracket satisfies $[(X, Y), (\tilde{X}, \tilde{Y})] = \mathrm{ad}_{(X,Y)}(X, Y)$, and the adjoint representation ad of the Lie algebra may be obtained from that of the Lie group by differentiating,

$$\mathrm{ad} = D(\mathrm{Ad})|_{(e_H, e_N)} . \tag{A.17}$$

In order to avoid the necessity of computing $\frac{\mathrm{d}}{\mathrm{d}t}\sigma_{n(t)}(\tilde{X})$ in this calculation, which would be quite cumbersome, we proceed as follows:

(a) Compute Lie brackets of the form $[(X, 0), (\tilde{X}, 0)]$, $[(0, Y), (0, \tilde{Y})]$, $[(X, 0), (0, Y)]$ by differentiating Ad as given by (A.12).
   Hint: For differentiating Ad, *consider curves in $H \ltimes_\rho N$ through the neutral element with specified derivative there, and use the chain rule, similar to the computation of* Ad *in the proof of* (A.12).
(b) Use your results from part (a) to express a general bracket. You should obtain the Lie bracket of $\mathfrak{h} \oplus_{\dot\rho'} \mathfrak{n}$.   △

---

[3] Of course, one could also compute the Lie bracket of a semidirect product group via the consideration of left-invariant vector fields, but the way via the adjoint representation is somewhat easier.

Semidirect Extensions of Principal Bundles

B

Abstract

Here, we will develop a natural construction that, given a semidirect product $H \ltimes N$ of two Lie groups, allows the extension of a principal bundle with structure group the non-normal subgroup H of the product to a larger bundle with structure group the product $H \ltimes N$, and classify connections on the extended bundle. This theory offers the natural global background for the description of Newton–Cartan gravity (and type I torsional Newton–Cartan gravity) in terms of the Bargmann group, as discussed in Chap. 5. (A sketch of the construction presented here is given in my (the author's) article [2].)

Construction B.1 Let H, N be Lie groups and $\rho\colon H \to \mathsf{Aut}(N)$ a group homomorphism such that the joint map $H \times N \ni (h, n) \mapsto \rho_h(n) \in N$ is smooth. Let $P \xrightarrow{\pi} M$ be a principal H-bundle. We can extend P to a principal $(H \ltimes N)$-bundle[4] $Q \xrightarrow{\hat{\pi}} M$ as follows: denoting by $\tilde{\rho}\colon H \to \mathsf{Diff}(H \ltimes N)$ the natural left action of H on $H \ltimes N$ by multiplication, i.e.

$$\tilde{\rho}_{h_2}(h_1, n) := (h_2, e_N)(h_1, n) = (h_2 h_1, \rho_{h_2}(n)), \tag{B.1a}$$

we define Q as the associated bundle

$$Q := P \times_{\tilde{\rho}} (H \ltimes N). \tag{B.1b}$$

The natural right action of $H \ltimes N$ on itself (by multiplication) induces a free right action on Q which is transitive on the fibres and compatible with the local trivialisations, thus making Q into a principal bundle as desired. Explicitly, the action is given by

$$[p, (h, n)] \cdot (\tilde{h}, \tilde{n}) := [p, (h, n)(\tilde{h}, \tilde{n})] = [p, (h\tilde{h}, n\rho_h(\tilde{n}))]. \tag{B.2}$$

---

[4] Here we mean of course the semidirect product with respect to ρ, omitting it from the notation.

P. K. Schwartz, *Newton–Cartan Gravity*, Lecture Notes in Physics 1044,
https://doi.org/10.1007/978-3-032-03967-5

157

We also obtain natural bundle homomorphisms $Q \overset{\beta}{\underset{\gamma}{\rightleftarrows}} P$ satisfying $\beta \circ \gamma = \mathrm{id}_P$, namely

$$\gamma(p) = [p, (e_H, e_N)], \quad \beta([p, (h, n)]) = ph . \tag{B.3}$$

By construction, with respect to local trivialisations $P \overset{\gamma}{\to} Q$ looks like the inclusion $H \hookrightarrow H \ltimes N$, such that it really exhibits Q as a principal bundle extension of P (i.e. P is a reduction of the structure group of Q from $H \ltimes N$ to H.) $\qquad \triangle$

We now want to classify connections on the extended bundle Q.

Lemma B.2 *We use the notation from Construction* B.1. *Let* $\hat{\omega} \in \Omega^1(Q, \mathfrak{h} \oplus \mathfrak{n})$ *be a connection on* Q. *We decompose its pullback along* γ *as* $\gamma^*\hat{\omega} = (\omega, \theta)$ *with* $\omega \in \Omega^1(P, \mathfrak{h})$ *and* $\theta \in \Omega^1(P, \mathfrak{n})$. *Then* ω *is a connection and* θ *is a* $\dot{\rho}$-*tensorial form on* P, *where* $\dot{\rho}: H \to \mathsf{Aut}(\mathfrak{n})$ *is the representation induced by* ρ.

Proof γ is H-equivariant, i.e. for any $h \in H$, we have $\gamma \circ R_h = R_{(h,e_N)} \circ \gamma$. Thus, the Ad-equivariance of $\hat{\omega}$ implies

$$\begin{aligned}
R_h^*(\gamma^*\hat{\omega}) &= (\gamma \circ R_h)^*\hat{\omega} \\
&= (R_{(h,e_N)} \circ \gamma)^*\hat{\omega} \\
&= \gamma^*(R_{(h,e_N)}^*\hat{\omega}) \\
&= \gamma^*(\mathrm{Ad}_{(h^{-1},e_N)} \circ \hat{\omega}) \\
&= \mathrm{Ad}_{(h^{-1},e_N)} \circ (\gamma^*\hat{\omega}).
\end{aligned} \tag{B.4}$$

Considering the form of the adjoint representation for a semidirect product (Proposition A.6), this means that

$$R_h^*\omega = \mathrm{Ad}_{h^{-1}} \circ \omega , \tag{B.5a}$$
$$R_h^*\theta = \dot{\rho}_{h^{-1}} \circ \theta . \tag{B.5b}$$

Now we consider the fundamental vector fields of the actions on the principal bundles. For $X \in \mathfrak{h}$, choosing a curve $h(t)$ in H with $\dot{h}(0) = X$, the equivariance of γ implies

$$\begin{aligned}
D\gamma(\tilde{X}|_p) &= D\gamma \left(\left. \frac{\mathrm{d}}{\mathrm{d}t} ph(t) \right|_{t=0} \right) \\
&= \left. \frac{\mathrm{d}}{\mathrm{d}t} \gamma(ph(t)) \right|_{t=0} \\
&= \left. \frac{\mathrm{d}}{\mathrm{d}t} \gamma(p) \cdot (h(t), e_N) \right|_{t=0} \\
&= \left. \widetilde{(X, 0)} \right|_{\gamma(p)} :
\end{aligned} \tag{B.6}$$

the equivariance of γ translates into $D\gamma \circ \tilde{X} = \widetilde{(X,0)} \circ \gamma$. Therefore, the condition $\hat{\omega}\left(\widetilde{(X,Y)}\big|_q\right) = (X,Y)$ for $(X,Y) \in \mathfrak{h} \oplus \mathfrak{n}$ implies

$$(\gamma^*\hat{\omega})(\tilde{X}|_p) = \hat{\omega}(D\gamma(\tilde{X}|_p)) = \hat{\omega}\left(\widetilde{(X,0)}\Big|_{\gamma(p)}\right) = (X,0). \tag{B.7}$$

Hence ω satisfies

$$\omega(\tilde{X}) = X, \tag{B.8}$$

and θ vanishes on vertical vectors. $\square$

In the proof of the next result, we need a small fact about Lie group actions:

Exercise B.3 (*Translation of fundamental vector fields*) Let G be a Lie group, with a smooth *right* action on a manifold M that we denote by

$$G \times M \ni (g,p) \mapsto pg = R_g(p) \in M. \tag{B.9}$$

For $X \in \mathfrak{g}$, we denote by $\tilde{X} \in \Gamma(TM)$ the corresponding fundamental vector field on M.

(a) Show that fundamental vector fields satisfy

$$(R_{g^{-1}})_*\tilde{X} = \widetilde{\mathrm{Ad}_g(X)}. \tag{B.10}$$

Hint: Explicitly write out the definition of the pushforward vector field evaluated at a point p, and use the definition of Ad_g as the differential of the conjugation map $\tilde{g} \mapsto g\tilde{g}g^{-1}$ at the neutral element.

(b) Show that the previous result implies

$$DR_g(\tilde{X}|_p) = \widetilde{\mathrm{Ad}_{g^{-1}}(X)}\Big|_{pg}. \tag{B.11}$$

(c) What would the corresponding equations look like for *left* actions? $\triangle$

Lemma B.4 *The above correspondence between connections on Q and pairs of connections and $\dot{\rho}$-tensorial one-forms on P is bijective, i.e. given a connection $\omega \in \Omega^1(P,\mathfrak{h})$ and a $\dot{\rho}$-tensorial $\theta \in \Omega^1_{\dot{\rho}}(P,\mathfrak{n})$, there is a unique connection $\hat{\omega} \in \Omega^1(Q,\mathfrak{h} \oplus \mathfrak{n})$ such that $\gamma^*\hat{\omega} = (\omega,\theta)$.*

Proof Let $q \in Q$, and write it in the (unique) form $q = [p,(e_H,n)]$. For $v \in T_qQ$, we want to define $\hat{\omega}(v)$. Since $q = [p,(e_H,e_N)] \cdot (e_H,n) = \gamma(p) \cdot (e_H,n)$, we need

$$\hat{\omega}|_q(v) = (R^*_{(e_H,n)}\hat{\omega})|_{\gamma(p)} \left(DR_{(e_H,n^{-1})}(v)\right)$$
$$= \mathrm{Ad}_{(e_H,n^{-1})}\left(\hat{\omega}|_{\gamma(p)}\left(DR_{(e_H,n^{-1})}(v)\right)\right). \tag{B.12}$$

Thus we need only consider the case $q = \gamma(p)$. So let $v \in T_{\gamma(p)}Q$. Such a v is not necessarily in the image of $D\gamma$, but since the 'new directions' in TQ (new compared to TP) are 'along N' in the fibres, v can be uniquely written as

$$v = D\gamma(\hat{v}) + \widetilde{(0, Y)}\Big|_{\gamma(p)} \tag{B.13a}$$

with $\hat{v} \in T_p P$ and $Y \in \mathfrak{n}$. In the case $\hat{v} = 0$, we need $\hat{\omega}(v) = (0, Y)$, and in the case $Y = 0$, we need $\hat{\omega}|_{\gamma(p)}(D\gamma(\hat{v})) = (\gamma^*\hat{\omega})|_p(\hat{v}) = (\omega|_p(\hat{v}), \theta|_p(\hat{v}))$. Thus $\hat{\omega}$ is required to be given by

$$\hat{\omega}|_{\gamma(p)}(v) = (\omega|_p(\hat{v}), \theta|_p(\hat{v}) + Y) \tag{B.13b}$$

on the image of γ.

We now have to check that the $\hat{\omega}$ thus defined—via (B.13) on the image of γ and extended to all of Q via (B.12)—is really a connection. Ad-equivariance by N holds by construction. Ad-equivariance by H on the image of γ can be seen as follows: for $v = D\gamma(\hat{v}) + \widetilde{(0, Y)}\Big|_{\gamma(p)} \in T_{\gamma(p)}Q$, we have

$$\mathrm{DR}_{(h,e_N)}(v) = \mathrm{DR}_{(h,e_N)}(D\gamma(\hat{v})) + \mathrm{DR}_{(h,e_N)}\left(\widetilde{(0, Y)}\Big|_{\gamma(p)}\right)$$

$$= D\gamma(\mathrm{DR}_h(\hat{v})) + \widetilde{\mathrm{Ad}_{(h^{-1},e_N)}(0, Y)}\Big|_{\gamma(p)\cdot(h,e_N)}$$

$$= D\gamma(\mathrm{DR}_h(\hat{v})) + \widetilde{\mathrm{Ad}_{(h^{-1},e_N)}(0, Y)}\Big|_{\gamma(ph)}$$

$$= D\gamma(\mathrm{DR}_h(\hat{v})) + \widetilde{(0, \dot{\rho}_{h^{-1}}(Y))}\Big|_{\gamma(ph)}, \tag{B.14}$$

where we used H-equivariance of γ and (B.11). Using this, we obtain

$$\begin{aligned}
(\mathrm{R}^*_{(h,e_N)}\hat{\omega})|_{\gamma(p)}(v) &= \hat{\omega}|_{\gamma(ph)}(\mathrm{DR}_{(h,e_N)}(v)) \\
&= \left(\omega|_{ph}(\mathrm{DR}_h(\hat{v})), \theta|_{ph}(\mathrm{DR}_h(\hat{v})) + \dot{\rho}_{h^{-1}}(Y)\right) \\
&= \left(\mathrm{Ad}_{h^{-1}}(\omega|_p(\hat{v})), \dot{\rho}_{h^{-1}}(\theta|_p(\hat{v}) + Y)\right) \\
&= \mathrm{Ad}_{(h^{-1},e_N)}(\hat{\omega}|_{\gamma(p)}(v)) \tag{B.15}
\end{aligned}$$

where we used the Ad-equivariance of ω and the $\dot{\rho}$-equivariance of θ. Combined, the Ad-equivariance of $\hat{\omega}$ on the whole of Q follows: for $q = \gamma(p) \cdot (e_H, n)$, we have

$$\begin{aligned}
(\mathrm{R}^*_{(h,e_N)}\hat{\omega})|_q &= \hat{\omega}|_{q\cdot(h,e_N)} \circ \mathrm{DR}_{(h,e_N)} \\
&= \hat{\omega}|_{\gamma(ph)\cdot(e_H, \rho_{h^{-1}}(n))} \circ \mathrm{DR}_{(h,e_N)} \\
\text{(B.12)} \quad &= \mathrm{Ad}_{(e_H, \rho_{h^{-1}}(n^{-1}))} \circ \hat{\omega}|_{\gamma(ph)} \circ \mathrm{DR}_{(e_H, \rho_{h^{-1}}(n^{-1}))} \circ \mathrm{DR}_{(h,e_N)}
\end{aligned}$$

$$
\begin{aligned}
&= \mathrm{Ad}_{(e_H,\,\rho_{h^{-1}}(n^{-1}))} \circ \hat{\boldsymbol{\omega}}|_{\gamma(ph)} \circ \mathrm{DR}_{(h,n^{-1})} \\
&= \mathrm{Ad}_{(e_H,\,\rho_{h^{-1}}(n^{-1}))} \circ \hat{\boldsymbol{\omega}}|_{\gamma(ph)} \circ \mathrm{DR}_{(h,e_N)} \circ \mathrm{DR}_{(e_H,n^{-1})} \\
\text{(B.12)}\quad &= \mathrm{Ad}_{(e_H,\,\rho_{h^{-1}}(n^{-1}))} \circ \mathrm{Ad}_{(h^{-1},e_N)} \circ \hat{\boldsymbol{\omega}}|_{\gamma(p)} \circ \mathrm{DR}_{(e_H,n^{-1})} \\
&= \mathrm{Ad}_{(h^{-1},e_N)} \circ \mathrm{Ad}_{(e_H,n^{-1})} \circ \hat{\boldsymbol{\omega}}|_{\gamma(p)} \circ \mathrm{DR}_{(e_H,n^{-1})} \\
\text{(B.15)}\quad &= \mathrm{Ad}_{(h^{-1},e_N)} \circ \hat{\boldsymbol{\omega}}|_q \; . \qquad\qquad\qquad\qquad \text{(B.16)}
\end{aligned}
$$

Finally, we have to check that $\hat{\boldsymbol{\omega}}$ take the correct value on fundamental vector fields. Since we already established Ad-equivariance, it is enough to show this on the image of γ. On fundamental vectors of the form $\widetilde{(0,Y)}\big|_{\gamma(p)}$, our form $\hat{\boldsymbol{\omega}}$ takes the value $(0,Y)$ by construction. For $\widetilde{(X,0)}\big|_{\gamma(p)}$, choosing a curve $h(t)$ in H such that $\dot{h}(0) = X$, we have

$$
\begin{aligned}
\hat{\boldsymbol{\omega}}\!\left(\widetilde{(X,0)}\Big|_{\gamma(p)}\right) &= \hat{\boldsymbol{\omega}}\left(\frac{\mathrm{d}}{\mathrm{d}t}\gamma(p)\cdot(h(t),e_N)\Big|_{t=0}\right) \\
&= \hat{\boldsymbol{\omega}}\left(\frac{\mathrm{d}}{\mathrm{d}t}\gamma(ph(t))\Big|_{t=0}\right) \\
&= \hat{\boldsymbol{\omega}}(\mathrm{D}\gamma(\tilde{X}|_p)) \\
\text{(B.13)}\quad &= (\omega(\tilde{X}|_p),\,\boldsymbol{\theta}(\tilde{X}|_p)) \\
&= (X,0). \qquad\qquad\qquad\qquad \text{(B.17)}
\end{aligned}
$$

This finishes the proof. $\qquad\qquad\qquad\qquad\qquad\qquad\qquad\qquad\qquad\qquad\quad\square$

Theorem B.5 *Let H,N be Lie groups, $\rho\colon H \to \mathsf{Aut}(N)$ a group homomorphism such that the joint map $H \times N \ni (h,n) \mapsto \rho_h(n) \in N$ is smooth, and $P \xrightarrow{\pi} M$ be a principal H-bundle. As in Construction B.1, let $\tilde{\rho}\colon H \to \mathsf{Diff}(H \ltimes N)$ be the natural left multiplication action of H on $H \ltimes N$, let $Q = P \times_{\tilde{\rho}} (H \ltimes N)$ the 'semidirect extension' of P via ρ, and $\gamma\colon P \to Q$ the natural embedding.*

Then connections $\hat{\boldsymbol{\omega}} \in \Omega^1(Q,\mathfrak{h}\oplus\mathfrak{n})$ on Q correspond bijectively to pairs of connections $\omega \in \Omega^1(P,\mathfrak{h})$ and $\dot{\rho}$-tensorial one-forms $\boldsymbol{\theta} \in \Omega^1_{\dot{\rho}}(P,\mathfrak{n})$ on P via the pullback condition

$$
\gamma^*\hat{\boldsymbol{\omega}} = (\omega,\boldsymbol{\theta}), \qquad\qquad\qquad\qquad \text{(B.18)}
$$

where $\dot{\rho}\colon H \to \mathsf{Aut}(\mathfrak{n})$ is the representation induced by ρ.

In this situation, the curvature form $\hat{\boldsymbol{R}} \in \Omega^2(Q,\mathfrak{h}\oplus\mathfrak{n})$ of $\hat{\boldsymbol{\omega}}$ satisfies

$$
\gamma^*\hat{\boldsymbol{R}} = (\boldsymbol{R},\, \mathrm{d}^\omega\boldsymbol{\theta} + \tfrac{1}{2}[\boldsymbol{\theta}\wedge\boldsymbol{\theta}]), \qquad\qquad \text{(B.19)}
$$

where $\boldsymbol{R} \in \Omega^2(p,\mathfrak{h})$ is the curvature form of ω.

Proof The bijective correspondence was established in Lemmas B.2 and B.4. Pulling back the structure equation $\hat{\boldsymbol{R}} = \mathrm{d}\hat{\boldsymbol{\omega}} + \frac{1}{2}[\hat{\boldsymbol{\omega}} \wedge \hat{\boldsymbol{\omega}}]$ with γ and using the explicit form of the semidirect sum Lie bracket, we obtain

$$
\begin{aligned}
\gamma^{*}\hat{\boldsymbol{R}} &= \mathrm{d}(\gamma^{*}\hat{\boldsymbol{\omega}}) + \tfrac{1}{2}[\gamma^{*}\hat{\boldsymbol{\omega}} \wedge \gamma^{*}\hat{\boldsymbol{\omega}}] \\
&= \mathrm{d}(\boldsymbol{\omega}, \boldsymbol{\theta}) + \tfrac{1}{2}[(\boldsymbol{\omega}, \boldsymbol{\theta}) \wedge (\boldsymbol{\omega}, \boldsymbol{\theta})] \\
&= (\mathrm{d}\boldsymbol{\omega}, \mathrm{d}\boldsymbol{\theta}) + \tfrac{1}{2}([\boldsymbol{\omega} \wedge \boldsymbol{\omega}], [\boldsymbol{\theta} \wedge \boldsymbol{\theta}] + 2\dot{\rho}'_{\boldsymbol{\omega}} \wedge \boldsymbol{\theta}) \\
&= (\boldsymbol{R}, \mathrm{d}^{\omega}\boldsymbol{\theta} + \tfrac{1}{2}[\boldsymbol{\theta} \wedge \boldsymbol{\theta}]).
\end{aligned}
\tag{B.20}
$$

$\square$

We can now apply this general theory to the classical example of the *affine frame bundle* of a manifold (see, e.g., ref. [1, Sect. III.3]).

Example B.6 We consider the above situation in the case that $P = F(M)$ is the linear frame bundle of an n-dimensional manifold, i.e. $H = \mathsf{GL}(n)$, and we let $N = \mathbb{R}^{n}$ and $\rho\colon \mathsf{GL}(n) \to \mathsf{Aut}(\mathbb{R}^{n})$ the defining representation; i.e. $H \ltimes N = \mathsf{GL}(n) \ltimes \mathbb{R}^{n}$ is the *affine group* in n dimensions. The extended bundle $Q = F(M) \times_{\mathsf{GL}(n)} (\mathsf{GL}(n) \ltimes \mathbb{R}^{n})$ from above is then naturally isomorphic to the *affine frame bundle* $\mathrm{Aff}(M)$ of M, whose fibres are given by

$$
\mathrm{Aff}_{x}(M) = F_{x}(M) \times T_{x}M,
\tag{B.21}
$$

i.e. whose elements are 'affine bases' of tangent spaces. The isomorphism is given by

$$
Q \ni [(e_{A}), (\mathbb{1}, (y^{A}))] \mapsto ((e_{A}), y^{B}e_{B}) \in \mathrm{Aff}(M)
\tag{B.22}
$$

A connection on $\mathrm{Aff}_{x}(M)$, or equivalently one on Q, is called a *generalised affine connection* on M. According to Theorem B.5, such a connection corresponds to a usual linear connection $\boldsymbol{\omega}$ on $F(M)$ and a tensorial form $\boldsymbol{\theta} \in \Omega^{1}_{\mathsf{GL}(n)}(F(M), \mathbb{R}^{n})$.

The tensorial form $\boldsymbol{\theta}$ corresponds to a form on the base manifold valued in an associated vector bundle,

$$
\theta \in \Omega^{1}(M, F(M) \times_{\mathsf{GL}(n)} \mathbb{R}^{n}),
\tag{B.23a}
$$

locally defined by

$$
\theta = [\sigma, \sigma^{*}\boldsymbol{\theta}]
\tag{B.23b}
$$

for local sections σ. If this θ is the canonical solder form of $F(M) \times_{\mathsf{GL}(n)} \mathbb{R}^{n}$, then the generalised affine connection we started with is called an *affine connection* on $\mathrm{Aff}(M)$.

For a generalised affine connection given by $(\boldsymbol{\omega}, \boldsymbol{\theta})$, according to Theorem B.5 the 'translational part' of its curvature is given by the exterior covariant derivative $\mathrm{d}^{\omega}\boldsymbol{\theta} \in \Omega^{2}_{\mathsf{GL}(n)}(F(M), \mathbb{R}^{n})$, which corresponds to a form $\mathrm{d}^{\omega}\theta \in \Omega^{2}(M, F(M) \times_{\mathsf{GL}(n)} \mathbb{R}^{n})$ on the base. In the case of an *affine* connection, i.e. if θ is the canonical solder form, according to Cartan's first structure equation this is the torsion of $\boldsymbol{\omega}$ (up to the identification $F(M) \times_{\mathsf{GL}(n)} \mathbb{R}^{n} \cong TM$ via θ). $\triangle$

References

1. Kobayashi, S., Nomizu, K.: Foundations of Differential Geometry, Volume I. Wiley Classics Library. John Wiley & Sons, New York (1996). Republication of original from 1963. https://www.wiley.com/en-us/Foundations+of+Differential+Geometry\%252C+Volume+1-p-9780471157335
2. Schwartz, P.K.: Teleparallel Newton–Cartan gravity. Class. Quantum Gravity **40**(10), 105008 (2023). https://doi.org/10.1088/1361-6382/accc02

Twistless-Torsional Newton–Cartan Expansion of the Lorentzian Geodesic Equation

C

Abstract

Here, we provide full details of the c^{-1} expansion of the Lorentzian geodesic equation in the situation of a generalised formal Newtonian $c \to \infty$ limit of a general-relativistic spacetime with $\tau \wedge d\tau = 0$, as discussed in Construction 6.7. As noted there, this expansion was developed by Hansen, Hartong, and Obers in the articles [1, 2].

Construction C.1 Let M, g, τ, h, m be as in Construction 6.7. This means the following: M is a manifold with a Lorentzian metric g that together with its inverse has formal c^{-1} expansions as in Lemma 3.48; further, the inverse metric $g^{-1} = h + c^{-2}m + \mathrm{O}(c^{-4})$ has no order-c^{-3} term, such that (6.13) holds. The clock form satisfies the TTNC condition $\tau \wedge d\tau = 0$, and we assume $d\tau$ to be nowhere-vanishing (since we are interested in the specifics of the case $d\tau \neq 0$).

The TTNC condition implies that we may write

$$d\tau = \tau \wedge \xi \tag{C.1}$$

for a one-form ξ (which is of course only unique up to addition of multiples of τ). This implies that the leading-order coefficient of the Christoffel symbols of g has the form

$$-\tau_{(\mu} (d\tau)_{\nu)}{}^{\rho} = -\tau_\mu \tau_\nu \xi^\rho . \tag{C.2}$$

Further, since $d\tau$ is nowhere-vanishing, ξ^ρ is nowhere-vanishing.

Let x be a curve in M, and y a vector field along x. We interpret these as arising by expansion in c^{-1} of a 'c-dependent curve' $X(\lambda)$ in M according to (6.16), i.e. according to

$$X^\mu(\lambda) = x^\mu(\lambda) + c^{-1}y^\mu(\lambda) + \mathrm{O}(c^{-2}). \tag{C.3}$$

In the following calculation, as in Construction 6.7 we will often omit the argument λ, and denote the derivative with respect to the argument λ by a prime.

© The Editor(s) (if applicable) and The Author(s), under exclusive license to Springer Nature Switzerland AG 2026

P. K. Schwartz, *Newton–Cartan Gravity*, Lecture Notes in Physics 1044,
https://doi.org/10.1007/978-3-032-03967-5

For a smooth function f on M, we define 'evaluation at X' by (6.17), i.e. by

$$f\big|_X = f\big|_x + c^{-1} y^\mu (\partial_\mu f)\big|_x + \mathrm{O}(c^{-2}) = f\big|_x + c^{-1}\mathrm{d}f\big|_x(y) + \mathrm{O}(c^{-2}). \quad \text{(C.4)}$$

This makes sense for any smooth function, so we may also apply it to the coordinate component functions of any geometric object on M (e.g. of a tensor field). In general, however, the second term in the expansion will then not directly have a coordinate-invariant interpretation; this will need some rewriting.

Using (C.4) to expand $\tau(X')$ in c^{-1}, we obtain

$$\begin{aligned}
\tau(X') &= \tau_\mu\big|_X X'^\mu \\
&= \big(\tau_\mu\big|_x + c^{-1} y^\nu (\partial_\nu \tau_\mu)\big|_x + \mathrm{O}(c^{-2})\big)\big(x'^\mu + c^{-1} y'^\mu + \mathrm{O}(c^{-2})\big) \\
&= \tau(x') + c^{-1}\big(\tau_\mu\big|_x y'^\mu + x'^\mu y^\nu (\partial_\nu \tau_\mu)\big|_x\big) + \mathrm{O}(c^{-2}) \\
&= \tau(x') + c^{-1} A + \mathrm{O}(c^{-2}),
\end{aligned} \quad \text{(C.5)}$$

where we defined

$$A(\lambda) := \tau_\mu\big|_{x(\lambda)} y'^\mu(\lambda) + x'^\mu(\lambda) y^\nu(\lambda)(\partial_\nu \tau_\mu)\big|_{x(\lambda)} . \quad \text{(C.6a)}$$

Note that the expression $\tau_\mu\big|_x y'^\mu$ does not have a coordinate-invariant meaning: y is a vector field along the curve x, so the derivatives $y'^\mu(\lambda)$ of its components are *not* the components of a vector in $T_{x(\lambda)}M$ to which $\tau\big|_{x(\lambda)}$ could be applied. However, using $\partial_\nu \tau_\mu = \partial_\mu \tau_\nu - (\mathrm{d}\tau)_{\mu\nu}$, we can rewrite the expansion in a manifestly coordinate-invariant way: this yields

$$\begin{aligned}
A &= \tau_\mu\big|_x y'^\mu + x'^\mu y^\nu (\partial_\mu \tau_\nu)\big|_x - \mathrm{d}\tau(x', y) \\
&= \tau_\mu\big|_x y'^\mu + y^\nu \frac{\mathrm{d}}{\mathrm{d}\lambda}(\tau_\nu\big|_x) - \mathrm{d}\tau(x', y) \\
&= \frac{\mathrm{d}}{\mathrm{d}\lambda}(\tau(y)) - \mathrm{d}\tau(x', y).
\end{aligned} \quad \text{(C.6b)}$$

Now we are going to expand the geodesic equation for X in powers of c^{-1}. Using the expansion (6.12) of the Christoffel symbols of g combined with (C.2), we obtain

$$\begin{aligned}
0 &= (\overset{g}{\nabla}_{X'} X')^\rho \\
&= X''^\rho + \overset{g}{\Gamma}{}^\rho_{\mu\nu}\big|_X X'^\mu X'^\nu \\
&= -c^2 \xi^\rho\big|_X (\tau(X'))^2 + (\overset{(0)}{\nabla}_{X'} X')^\rho + \mathrm{O}(c^{-1}),
\end{aligned} \quad \text{(C.7)}$$

where $\overset{(0)}{\nabla}$ is the connection that we obtain as the order-c^0 term of $\overset{g}{\nabla}$. Explicitly, by (6.12), this is given by

$$\overset{(0)}{\Gamma}{}^\rho_{\mu\nu} = m^{\rho\sigma}\tau_{(\mu}(\mathrm{d}\tau)_{\nu)\sigma} + \bar{\Gamma}^\rho_{\mu\nu} - \frac{1}{2}\hat{v}^\rho(\mathrm{d}\tau)_{\mu\nu} . \quad \text{(C.8)}$$

Using the expansion (C.5) of $\tau(X')$, we see that the leading-order term of the geodesic equation (C.7) is the order-c^2 term, yielding $0 = \xi^\rho|_x(\tau(x'))^2$. Since we assumed $\xi^\rho \neq 0$, this is equivalent to

$$\tau(x') = 0, \tag{C.9}$$

which is the first result stated in Construction 6.7 (Eq. (6.18a)).

Assuming that this holds, inserting (C.5) into the expanded geodesic equation (C.7) we obtain

$$0 = -\xi^\rho\big|_x A^2 + (\overset{(0)}{\nabla}_{X'} X')^\rho + \mathrm{O}(c^{-1}). \tag{C.10a}$$

Expanding $\overset{(0)}{\Gamma}{}^\rho_{\mu\nu}\big|_X = \overset{(0)}{\Gamma}{}^\rho_{\mu\nu}\big|_x + \mathrm{O}(c^{-1})$ according to (C.4), this becomes

$$0 = -\xi^\rho\big|_x A^2 + (\overset{(0)}{\nabla}_{x'} x')^\rho + \mathrm{O}(c^{-1}). \tag{C.10b}$$

Hence, the order-c^0 term of the geodesic equation is

$$\overset{(0)}{\nabla}_{x'} x' = A^2 h(\xi, \cdot)\big|_x . \tag{C.11a}$$

Since x is a spacelike curve due to the order-c^2 equation (C.7), we can replace $\overset{(0)}{\nabla}$ by the connection induced on the spatial leaf Σ on which x lies. Restricting the explicit form (C.8) to spacelike fields, using that $d\tau = \tau \wedge \xi$, we see that this restriction of $\overset{(0)}{\nabla}$ agrees with that of $\bar{\nabla}$. Now since $\bar{\nabla}$ is a Galilei connection, its restriction to the spatial leaf Σ is compatible with the induced metric $^{(n)}h$. Further, due to the torsion of $\bar{\nabla}$ being $\hat{v} \otimes d\tau = \hat{v} \otimes (\tau \wedge \xi)$, the restriction of $\bar{\nabla}$ to Σ is torsion-free. Therefore, the restriction is actually the Levi-Civita connection $\overset{(n)}{\nabla}$ of $(\Sigma, {}^{(n)}h)$. Thus, the equation (C.11a) for x can be rewritten as

$$\overset{(n)}{\nabla}_{x'} x' = A^2 h(\xi, \cdot)\big|_x . \tag{C.11b}$$

In this equation, the non-uniqueness of ξ does not play a role: ξ is unique up to addition of a multiple of τ, which leaves (C.11b) invariant.

Note further that due to the presence of A, (C.11b) is not an equation solely for x—it is coupled to y. In the following, we are going to show how (part of) the next order of the geodesic equation gives rise to an equation for y, and how this allows to decouple (C.11b) from y.

Contracting the geodesic equation for X with τ, by (6.13) and $d\tau = \tau \wedge \xi$ we obtain

$$0 = \tau_\rho\big|_X (\overset{g}{\nabla}_{X'} X')^\rho$$
$$= \tau_\mu\big|_X X''^\mu + (\tau_\rho \overset{g}{\Gamma}{}^\rho_{\mu\nu})\big|_X X'^\mu X'^\nu$$

$$
\begin{aligned}
&= \tau_\mu\big|_X X''^\mu + \big(\xi(\hat{v})\tau_\mu\tau_\nu - \tau_{(\mu}\xi_{\nu)} + \partial_{(\mu}\tau_{\nu)} + O(c^{-2})\big)\big|_X X'^\mu X'^\nu \\
&= \tau_\mu\big|_X X''^\mu + \xi(\hat{v})\big|_X (\tau(X'))^2 - \tau(X')\xi(X') + (\partial_\mu\tau_\nu)\big|_X X'^\mu X'^\nu + O(c^{-2}).
\end{aligned}
$$

$$\text{(C.12a)}$$

Using (C.4), $\tau(X') = c^{-1}A + O(c^{-2})$, and $\xi(X') = \xi(x') + O(c^{-1})$, we may further expand this expression in c^{-1}, yielding

$$
\begin{aligned}
0 &= \tau_\rho\big|_X (\overset{g}{\nabla}_{X'} X')^\rho \\
&= \big(\tau_\mu\big|_x + c^{-1}y^\nu(\partial_\nu\tau_\mu)\big|_x + O(c^{-2})\big)\big(x''^\mu + c^{-1}y''^\mu + O(c^{-2})\big) - c^{-1}A\xi(x') \\
&\quad + (\partial_\mu\tau_\nu)\big|_X \big(x'^\mu + c^{-1}y'^\mu + O(c^{-2})\big)\big(x'^\nu + c^{-1}y'^\nu + O(c^{-2})\big) + O(c^{-2}) \\
&= (c^0 \text{ terms}) + c^{-1}\big(\tau_\mu\big|_x y''^\mu + y^\nu x''^\mu(\partial_\nu\tau_\mu)\big|_x - A\xi(x') \\
&\quad + y^\kappa(\partial_\kappa\partial_\mu\tau_\nu)\big|_x x'^\mu x'^\nu + 2(\partial_{(\mu}\tau_{\nu)})\big|_x x'^\mu y'^\nu\big) + O(c^{-2}).
\end{aligned}
$$

$$\text{(C.12b)}$$

The order-c^0 part of this equation is the c^0 part of the geodesic equation contracted with $\tau\big|_x$. Therefore, assuming that the order-c^0 geodesic equation holds (i.e. (C.11b) is satisfied), the c^0 coefficient vanishes. For rewriting the c^{-1} part, we observe that

$$
\begin{aligned}
\frac{\mathrm{d}}{\mathrm{d}\lambda}A &= \frac{\mathrm{d}}{\mathrm{d}\lambda}\big(\tau_\mu\big|_x y'^\mu + x'^\mu y^\nu(\partial_\nu\tau_\mu)\big|_x\big) \\
&= \frac{\mathrm{d}}{\mathrm{d}\lambda}(\tau_\mu\big|_x)y'^\mu + \tau_\mu\big|_x y''^\mu + x''^\mu y^\nu(\partial_\nu\tau_\mu)\big|_x \\
&\quad + x'^\mu y'^\nu(\partial_\nu\tau_\mu)\big|_x + x'^\mu y^\nu\frac{\mathrm{d}}{\mathrm{d}\lambda}\big((\partial_\nu\tau_\mu)\big|_x\big) \\
&= x'^\kappa(\partial_\kappa\tau_\mu)\big|_x y'^\mu + \tau_\mu\big|_x y''^\mu + x''^\mu y^\nu(\partial_\nu\tau_\mu)\big|_x \\
&\quad + x'^\mu y'^\nu(\partial_\nu\tau_\mu)\big|_x + x'^\mu y^\nu x'^\kappa(\partial_\kappa\partial_\nu\tau_\mu)\big|_x \\
&= \tau_\mu\big|_x y''^\mu + x''^\mu y^\nu(\partial_\nu\tau_\mu)\big|_x + 2x'^\mu y'^\nu(\partial_{(\mu}\tau_{\nu)})\big|_x + x'^\mu y^\nu x'^\kappa(\partial_\nu\partial_\kappa\tau_\mu)\big|_x .
\end{aligned}
$$

$$\text{(C.13)}$$

This is precisely the c^{-1} coefficient of (C.12b), apart from the $-A\xi(x')$ term. Hence, this order of the equation is equivalent to

$$
0 = \frac{\mathrm{d}}{\mathrm{d}\lambda}A - A\xi(x'). \tag{C.14}
$$

Note that similar to the Eq. (C.11b) for x, this equation is insensitive to the non-uniqueness of ξ (addition of multiples of τ) due to $\tau(x') = 0$.

We are now going to show how (C.14) may be used to decouple the Eq. (C.11b) that we obtained for x from y. For this, we use that due to the TTNC condition $\tau \wedge \mathrm{d}\tau = 0$, we can locally write

$$
\tau = N\mathrm{d}T \tag{C.15}
$$

for smooth functions T and N. The exterior derivative of τ is then given by $d\tau = dN \wedge dT = \tau \wedge (-N^{-1}dN)$, such that for the one-form ξ characterising $d\tau$ we may choose

$$\xi = -N^{-1}dN. \tag{C.16}$$

We will use this to reformulate the equations obtained from the expansion of the geodesic equation in terms of N. (The non-uniqueness of N and its consequences for the resulting equations are discussed in Construction 6.7.)

Equation (C.16) implies that $\xi(x') = -N^{-1}\big|_x \frac{d}{d\lambda}N\big|_x$. Using this, the differential equation (C.14) we obtained for A becomes

$$0 = \frac{d}{d\lambda}A + AN^{-1}\big|_x \frac{d}{d\lambda}N\big|_x \,, \tag{C.17a}$$

or equivalently

$$
\begin{aligned}
0 &= N\big|_x \frac{d}{d\lambda}A + A\frac{d}{d\lambda}N\big|_x \\
&= \frac{d}{d\lambda}(A \cdot N\big|_x).
\end{aligned}
\tag{C.17b}
$$

This means that

$$A = C \cdot N^{-1}\big|_x \tag{C.17c}$$

for some integration constant $C \in \mathbb{R}$.

Inserting (C.17c) into the Eq. (C.11b) for x, and expressing ξ in terms of N according to (C.16), we obtain

$$
\begin{aligned}
\overset{(n)}{\nabla}_{x'}x' &= C^2 h(N^{-2}\xi, \cdot)\big|_x \\
&= -C^2 h(N^{-3}dN, \cdot)\big|_x \\
&= \frac{1}{2}C^2 h(dN^{-2}, \cdot)\big|_x \,.
\end{aligned}
\tag{C.18}
$$

Thus we have indeed decoupled the equation for x from y. This equation is the second result stated in Construction 6.7 (Eq. (6.18b)).

Rewritten in terms of N, the manifestly coordinate-invariant expression (C.6b) for A becomes

$$
\begin{aligned}
A &= \frac{d}{d\lambda}(\tau(y)) - d\tau(x', y) \\
&= \frac{d}{d\lambda}(\tau(y)) + \tau(y)\xi(x') \\
&= \frac{d}{d\lambda}(\tau(y)) - \tau(y)N^{-1}\big|_x \frac{d}{d\lambda}N\big|_x \,.
\end{aligned}
\tag{C.19}
$$

Equation (C.17c) for A, which we obtained by integrating the order-c^{-1} coefficient of the geodesic equation contracted with τ, thus becomes an equation for $\tau(y)$, namely

$$\frac{\mathrm{d}}{\mathrm{d}\lambda}(\tau(y)) - \tau(y)N^{-1}\big|_x \frac{\mathrm{d}}{\mathrm{d}\lambda}N\big|_x = C \cdot N^{-1}\big|_x. \tag{C.20}$$

This is the third result stated in Construction 6.7 (Eq. (6.18c)).

Next, expanding $g(X', X')$ in c^{-1}, we obtain

$$\begin{aligned}
g(X', X') &= -c^2(\tau(X'))^2 + \underset{\hat{v}}{h}(X', X') - 2\hat{\phi}\big|_X(\tau(X'))^2 + \mathrm{O}(c^{-2}) \\
&= -A^2 + \underset{\hat{v}}{h}(x', x') + \mathrm{O}(c^{-2}) \\
&= -A^2 + {}^{(n)}h(x', x') + \mathrm{O}(c^{-2}),
\end{aligned} \tag{C.21}$$

where we used the expansion (C.5) of $\tau(X')$, and the leading-order equation $\tau(x') = 0$. Writing the potential of the effective particle motion (C.18) as $V = -\frac{1}{2}C^2 \cdot N^{-2} = -\frac{1}{2}A^2$, this may be rewritten as

$$\frac{1}{2}g(X', X') = \frac{1}{2}{}^{(n)}h(x', x') + V + \mathrm{O}(c^{-2}), \tag{C.22}$$

reproducing Eq. (6.19) from Construction 6.7.

Finally, we are going to compute the velocity of the particle with Lorentzian worldline X, as measured by observers moving along $V = -g^{-1}(\tau, \cdot)$. In general, given two Lorentzian timelike vectors $w, \tilde{w}$ pointing in the same time direction (i.e. $g(w, \tilde{w}) < 0$), the velocity of an object whose motion is described by $\tilde{w}$ as measured by an observer along w is an element of the g-orthogonal complement of w, given by

$$c\left(\sqrt{-g(w, w)} \cdot \frac{\tilde{w}}{-g(w, \tilde{w})} - \frac{w}{\sqrt{-g(w, w)}}\right). \tag{C.23}$$

In particular, using

$$\begin{aligned}
V &= -g^{-1}(\tau, \cdot) \\
&= -(h + c^{-2}m + \mathrm{O}(c^{-4}))(\tau, \cdot) \\
&= -c^{-2}m(\tau, \cdot) + \mathrm{O}(c^{-4}) \\
&= -c^{-2}\hat{v} + \mathrm{O}(c^{-4})
\end{aligned} \tag{C.24}$$

and $\sqrt{-g^{-1}(\tau, \tau)} = c^{-1} + \mathrm{O}(c^{-2})$, we obtain that the velocity of $\tilde{w}$ with respect to V is

$$c\left(\sqrt{-g^{-1}(\tau, \tau)} \cdot \frac{\tilde{w}}{\tau(\tilde{w})} - \frac{w}{\sqrt{-g^{-1}(\tau, \tau)}}\right) = (1 + \mathrm{O}(c^{-1}))\frac{\tilde{w}}{\tau(\tilde{w})} - \hat{v} + \mathrm{O}(c^{-1}). \tag{C.25}$$

Applying this to the free test particle whose worldline is our geodesic X, its velocity with respect to V is

$$(1 + \mathrm{O}(c^{-1})) \frac{X'}{\tau(X')} - \hat{v} + \mathrm{O}(c^{-1}) = c\,\frac{x'}{A} + \mathrm{O}(c^0) = c\,\frac{x'}{C \cdot N^{-1}\big|_x} + \mathrm{O}(c^0),$$

$$(\mathrm{C}.26)$$

where we used the expansion (C.5) of $\tau(X')$, $\tau(x') = 0$, and (C.17c). Thus we have reproduced (6.20), the final result from Construction 6.7.   △

References

1. Hansen, D., Hartong, J., Obers, N.A.: Gravity between Newton and Einstein. Int. J. Mod. Phys. D **28**(14), 1944010 (2019). https://doi.org/10.1142/S0218271819440103
2. Hansen, D., Hartong, J., Obers, N.A.: Non-relativistic gravity and its coupling to matter. J. High Energy Phys. **2020**(6), 145 (2020). https://doi.org/10.1007/JHEP06(2020)145

Index

P. K. Schwartz, *Newton–Cartan Gravity*, Lecture Notes in Physics 1044,
https://doi.org/10.1007/978-3-032-03967-5

<u>GPSR Compliance</u>

The European Union's (EU) General Product Safety Regulation (GPSR) is a set of rules that requires consumer products to be safe and our obligations to ensure this.

If you have any concerns about our products, you can contact us on ProductSafety@springernature.com

In case Publisher is established outside the EU, the EU authorized representative is:

Springer Nature Customer Service Center GmbH
Europaplatz 3
69115 Heidelberg, Germany

Batch number: 09560227

Printed by Printforce, the Netherlands